Wildfire Hazards, Risks, and Disasters

T0320818

Hazards and Disasters Series

Wildfire Hazards, Risks, and Disasters

Volume Editor

Douglas Paton
School of Medicine (Psychology)
University of Tasmania
Launceston
Tasmania
Australia

Associate Editors

Petra T. Buergelt, Sarah McCaffrey and Fantina Tedim

Series Editor

John F. Shroder
Emeritus Professor of Geography and Geology
Department of Geography and Geology
University of Nebraska at Omaha
Omaha, NE

ELSEVIER

AMSTERDAM • BOSTON • HEIDELBERG • LONDON • NEW YORK • OXFORD
PARIS • SAN DIEGO • SAN FRANCISCO • SINGAPORE • SYDNEY • TOKYO

Elsevier
Radarweg 29, PO Box 211, 1000 AE Amsterdam, Netherlands
The Boulevard, Langford Lane, Kidlington, Oxford OX5 1GB, UK
225 Wyman Street, Waltham, MA 02451, USA

Notices
Knowledge and best practice in this field are constantly changing. As new research and experience broaden our understanding, changes in research methods, professional practices, or medical treatment may become necessary.

Practitioners and researchers must always rely on their own experience and knowledge in evaluating and using any information, methods, compounds, or experiments described herein. In using such information or methods they should be mindful of their own safety and the safety of others, including parties for whom they have a professional responsibility.

To the fullest extent of the law, neither the Publisher nor the authors, contributors, or editors, assume any liability for any injury and/or damage to persons or property as a matter of products liability, negligence or otherwise, or from any use or operation of any methods, products, instructions, or ideas contained in the material herein.

Library of Congress Cataloging-in-Publication Data
Application submitted

British Library Cataloguing in Publication Data
A catalogue record for this book is available from the British Library

ISBN: 978-0-12-410434-1

For information on all Elsevier publications
visit our web site at http://store.elsevier.com

This book has been manufactured using Print on Demand technology. Each copy is produced to order and is limited to black ink. The online version of this book will show color figures where appropriate.

Working together
to grow libraries in
developing countries

www.elsevier.com • www.bookaid.org

Contents

1. **Wildfires: International Perspectives on Their Social–Ecological Implications**

    Douglas Paton, Petra T. Buergelt, Fantina Tedim and Sarah McCaffrey

2. **Social Science Findings in the United States**

    Sarah McCaffrey, Eric Toman, Melanie Stidham and Bruce Shindler

3. **Wildfire: A Canadian Perspective**

    Tara McGee, Bonita McFarlane and Cordy Tymstra

8. Discourse on Taiwanese Forest Fires

Jan-Chang Chen and Chaur-Tzuhn Chen

9. Wildfires in India: Tools and Hazards

Joachim Schmerbeck and Daniel Kraus

10. System of Wildfires Monitoring in Russia

Evgeni I. Ponomarev, Valeri Ivanov and Nikolay Korshunov

11. Wildland Fire Danger Rating and Early Warning Systems

William J. de Groot, B. Michael Wotton and Michael D. Flannigan

Contributors

J. Antonio Alloza, CEAM, Parque Tecnológico, Ch. Darwin 14, Paterna, Spain

Petra T. Buergelt, Charles Darwin University, School of Psychological & Clinical Sciences, Darwin, Australia, University of Western Australia, Centre for Social Impact and Oceans Institute, University of Western Australia, Australia & Joint Centre for Disaster Research, Massey University, Mt Cook, Wellington, New Zealand

Jan-Chang Chen, Assistant Professor, Department of Forestry, National Pingtung University of Science and Technology, Pingtung, Taiwan

Chaur-Tzuhn Chen, Professor, Department of Forestry, National Pingtung University of Science and Technology, Pingtung, Taiwan

William J. de Groot, Natural Resources Canada − Canadian Forest Service, Sault Ste. Marie, ON, Canada

Michael D. Flannigan, Dept. of Renewable Resources, University of Alberta, Edmonton, AB, Canada

Michael Flannigan, Faculty of Forestry, University of Toronto, Toronto, ON, Canada

Valeri Ivanov, Siberian State Technological University, pr. Mira, Krasnoyarsk, Russia

Guillermo Julio-Alvear, Forest Fire Laboratory, University of Chile, Santiago, Chile

Nikolay Korshunov, Russian Institute of Continuous Education in Forestry, Pushkino, Moscow obl., Russia

Daniel Kraus, European Forest Institute (EFI), EFICENT Regional Office, Freiburg, Germany

Vittorio Leone, University of Basilicata (retired), Department of Crop Systems, Forestry and Environmental Sciences, Potenza, Italy

Sarah McCaffrey, Northern Research Station, USDA Forest Service, Evanston, IL, USA

Bonita McFarlane, Natural Resources Canada, Canadian Forest Service, Northern Forestry Centre, Edmonton, AB, Canada

Tara McGee, Department of Earth and Atmospheric Sciences, University of Alberta, Edmonton, AB, Canada

Douglas Paton, School of Medicine (Psychology), University of Tasmania, Launceston, Tasmania, Australia

Evgeni I. Ponomarev, V.N. Sukachev Institute of Forest, Siberian Branch of Russian Academy of Sciences, Akademgorodok, Krasnoyarsk, Russia; Siberian Federal University, pr. Svobodnyi, Krasnoyarsk, Russia

V. Ramon Vallejo, CEAM, Parque Tecnológico, Ch. Darwin 14, Paterna, Spain; Dept. Biologia Vegetal, Universitat de Barcelona, Diagonal 643, Barcelona, Spain

Saut Sagala, School of Architecture, Planning, and Policy Development, ITB, Indonesia

Roberto Garfias Salinas, Forest Fire Laboratory, University of Chile, Santiago, Chile

Joachim Schmerbeck, TERI University Department of Natural Resources, Vasant Kunj, New Delhi, India

Bruce Shindler, Department of Forest Ecosystems and Society, Oregon State University, Corvallis, OR, USA

Efraim Sitinjak, Resilience Development Initiative, Bandung, Indonesia

Ralph Smith, Department of Fire & Emergency Services, Cockburn Central, Western Australia, Australia

Miguel Castillo Soto, Forest Fire Laboratory, University of Chile, Santiago, Chile

Melanie Stidham, School of Environment and Natural Resources, The Ohio State University, Columbus, OH, USA

Fantina Tedim, Faculty of Arts, University of Porto, Geography Department, Porto, Portugal

Eric Toman, School of Environment and Natural Resources, The Ohio State University, Columbus, OH, USA

Cordy Tymstra, Alberta Environment and Sustainable Resource Development, Forestry and Emergency Response Division, Wildfire Management Branch, Edmonton, AB, Canada

B. Michael Wotton, Faculty of Forestry, University of Toronto, Toronto, ON, Canada

Gavriil Xanthopoulos, Hellenic Agricultural Organization "Demeter", Institute of Mediterranean Forest Ecosystems and Forest Products Technology, Athens, Greece

Dodon Yamin, Resilience Development Initiative, Bandung, Indonesia

General hazards, risks, and disasters: In general, hazards are processes that produce danger to human life and infrastructure. Risks are the potential or possibilities that something bad will happen because of the hazards. Disasters are that quite unpleasant result of the hazard occurrence that caused destruction of lives and infrastructure. Hazards, risks, and disasters have been coming under increasing strong scientific scrutiny in recent decades as a result of a combination of numerous unfortunate factors, many of which are quite out of control as a result of human actions. At the top of the list of exacerbating factors to any hazard, of course, is the tragic exponential population growth that is clearly not possible to maintain indefinitely on a finite Earth. As our planet is covered ever more with humans, any natural or human-caused (unnatural) hazardous process is increasingly likely to adversely impact life and construction systems. The volumes on hazards, risks, and disasters that we present here are thus an attempt to increase understandings about how to best deal with these problems, even while we all recognize the inherent difficulties of even slowing down the rates of such processes as other compounding situations spiral on out of control, such as exploding population growth and rampant environmental degradation.

Some natural hazardous processes such as volcanoes and earthquakes that emanate from deep within the Earth's interior are in no way affected by human actions, but a number of others are closely related to factors affected or controlled by humanity, even if however unwitting. Chief among these, of course, are climate-controlling factors, and no small measure of these can be exacerbated by the now obvious ongoing climate change at hand (Hay, 2013). Pervasive range fires and forest fires caused by human-enhanced or -induced droughts and fuel loadings, megaflooding into sprawling urban complexes on floodplains and coastal cities, biological threats from locust plagues, and other ecological disasters gone awry; all of these and many others are but a small part of the potentials for catastrophic risk that loom at many different scales, from the local to planet girdling.

In fact, the denial of possible planetwide catastrophic risk (Rees, 2013) as exaggerated jeremiads in media landscapes saturated with sensational science stories and end-of-the-world, Hollywood productions is perhaps quite understandable, even if simplistically shortsighted. The "end-of-days" tropes promoted by the shaggy-minded prophets of doom have been with us for centuries, mainly because of Biblical verse written in the early Iron Age during remarkably pacific times of only limited environmental change. Nowadays, however, the

Armageddon enthusiasts appear to want the worst to validate their death desires and prove their holy books. Unfortunately we are all entering times when just a few individuals could actually trigger societal breakdown by error or terror, if Mother Nature does not do it for us first. Thus we enter contemporaneous times of considerable peril that present needs for close attention.

These volumes we address here about hazards, risks, and disasters are not exhaustive dissertations about all the dangerous possibilities faced by the ever-burgeoning human populations, but they do address the more common natural perils that people face, even while we leave aside (for now) the thinking about higher-level existential threats from such things as bio- or cybertechnologies, artificial intelligence gone awry, ecological collapse, or runaway climate catastrophes.

In contemplating existential risk (Rossbacher, 2013), we have lately come to realize that the new existentialist philosophy is no longer the old sense of disorientation or confusion at the apparently meaninglessness or hopelessly absurd worlds of the past, but instead an increasing realization that serious changes by humans appear to be afoot that even threaten all life on the planet (Kolbert, 2014; Newitz, 2013). In the geological times of the Late Cretaceous, an asteroid collision with Earth wiped out the dinosaurs and much other life; at the present time, in contrast, humanity itself appears to be the asteroid.

Misanthropic viewpoints aside, however, an increased understanding of all levels and types of the more common natural hazards would seem a useful endeavor to enhance knowledge accessibility, even while we attempt to figure out how to extract ourselves and other life from the perils produced by the strong climate change so obviously underway. Our intent in these volumes is to show the latest good thinking about the more common endogenetic and exogenetic processes and their roles as threats to everyday human existence. In this fashion, the chapter authors and volume editors have undertaken to show you overviews and more focused assessments of many of the chief obvious threats at hand that have been repeatedly shown on screen and print media in recent years. As this century develops, we may come to wish that these examples of hazards, risks, and disasters are not somehow eclipsed by truly existential threats of a more pervasive nature. The future always hangs in the balance of opposing forces; the ever-lurking, but mindless threats from an implacable nature, or the heedless bureaucracies countered only sometimes in small ways by the clumsy and often feeble attempts by individual humans to improve our little lots in life. Only through improved education and understanding will any of us have a chance against such strong odds; perhaps these volumes will add some small measure of assistance in this regard.

Wildfire hazards, risks, and disasters: As a specific hazard, wildfires in rangeland, bushlands, taiga-tundra, and temperate and tropical forests world-wide are a great threat, especially where humans have built homes out into them, as well as in these times of recurrent drought brought upon by climate change. The results can be the horrendous near-annual fire events in places such as the Australian bushlands, in the southern California maquis (shrub-land) vegetation in the vast suburbs, and wherever else in the world a region is extensively built up by human constructions into a seasonably flammable environment. As human populations increase inexorably worldwide and the need for evermore land upon which to live leads directly to encroachment into areas that have burned repeatedly in the past, so this wildfire hazard increasingly looms ever larger to more and more people.

As Mockenhaupt (2014) has observed, the firefighters' conundrum is how best to balance risk in the growing wildland—urban interface. Faced as we are in the natural hazard landscape with volcanoes, floods, tsunamis, or hurricanes, we can do little but to let nature run its course, try to limit the damage, and clean up the aftermath. But when it comes to wildfire, we think that we can fight it, rather than allowing small fires back into the landscape as they have existed forever until natural human hubris imagined that wildfire could be banished forever. Young firefighters die every year, even though we realize that fire suppression is a battle that can never really be won, and that in some cases should not even be fought. Instead our valiant firefighters battle ever-larger wildfires to protect increasing numbers of homes. The result is a cycle of tragic inevitability.

Because wildfire hazards have apparently increased so much in recent years as the combined results of the ever-larger droughts, increased vulnerable populations, and more pervasive reporting, so too have been the increased number of remedial methods in dealing with fires. Volumes on prevention and warning, remediation and revegetation, establishment of new procedures and assessment processes provide assessments of fire hazard in many geographic areas across the world. If people are concerned about the area in which they live, a reading of this volume will at least give them enough basic information and ideas germane to assessing their own situation, or in seeking to learn more. The result might be an increase in their own factors of safety, which is a potentially valuable addition to knowledge thereby. The reader is encouraged to seek greater knowledge of their own local fire hazards.

John (Jack) Shroder
Editor-in-Chief
July 14, 2014

REFERENCES

Hay, W.W., 2013. Experimenting on a Small Planet: A Scholarly Entertainment. Springer-Verlag, Berlin, 983 p.

Kolbert, E., 2014. The Sixth Extinction: An Unnatural History. Henry Holt & Company, NY, 319 p.

Mockenhaupt, B., June 2014. Fire on the mountain. The Atlantic, 72—86.

Newitz, A., 2013. Scatter, Adapt, and Remember. Doubleday, NY, 305 p.

Rees, M., 2013. Denial of catastrophic risks. Science 339 (6124), 1123.

Rossbacher, L.A., October 2013. Contemplating existential risk. Earth, Geologic Column 58 (10), 64.

Chapter 1

Wildfires: International Perspectives on Their Social–Ecological Implications

Douglas Paton
School of Medicine (Psychology), University of Tasmania, Launceston, Tasmania, Australia

Petra T. Buergelt
Charles Darwin University, School of Psychological & Clinical Sciences, Darwin, Australia, University of Western Australia, Centre for Social Impact and Oceans Institute, University of Western Australia, Australia & Joint Centre for Disaster Research, Massey University, Mt Cook, Wellington, New Zealand

Fantina Tedim
Faculty of Arts, University of Porto, Geography Department, Porto, Portugal

Sarah McCaffrey
Northern Research Station, USDA Forest Service, Evanston, IL, USA

ABSTRACT

This chapter introduces the fact that of the several natural hazards contemporary communities may encounter, the complex interdependencies that exist between people and the forest sources of wildfire hazards make wildfire a unique hazard. It then proceeds to provide an overview of how historical patterns of interdependence between people and forests coupled with recent trends in population growth and their encroachment on forest environments for lifestyle and recreation are increasing risk. Next, it outlines a social–ecological approach to framing and managing wildfire risk and discusses environmental, ecological, and social factors that play complementary roles in the development and thus the management of wildfire risk. The chapter introduces international case studies that discuss the historical, social, cultural, and ecological aspects of wildfire risk management in countries with a long history of dealing with this hazard (e.g., United States and Australia) and in countries (e.g., Taiwan) where wildfire hazards represent a new and growing threat to the social and ecological landscape.

1.1 INTRODUCTION

Among the various natural hazards that contemporary communities may have to contend with, wildfires are unique. Wildfire attains this unique characteristic

Wildfire Hazards, Risks, and Disasters. http://dx.doi.org/10.1016/B978-0-12-410434-1.00001-4

as a result of the complex interdependencies that exist between people and the forest sources of wildfire hazards. For millennia, forests have delivered fuel and building materials to people, provided contexts for agricultural, livelihood and hunting, and they are increasingly sought out for their lifestyle and amenity values as people choose to live in close proximity to forests or use them for recreational purposes. Further, forest and other natural environments also play crucial roles in sustaining human well-being (Clayton and Opotow, 2003). These factors have influenced how people think about and act toward the source of the hazard. This relationship has also resulted in people's actions, increasing their risk. Consequently, people have a relationship with the source of the hazard that is not generally shared with any other natural hazard.

This long-standing interdependence between people and forests (as amenity and hazard) has resulted in many populations, particularly indigenous peoples, developing sustainable ways of relating to forests and the hazards they periodically create. Fire has been a prominent part of ecosystems for millennia, and ecosystems have evolved through people working with fire rather than against it (see Chapters 2, 3, 5, and 7). Thus, a significant point of departure between wildfire and other natural hazards is that fire serves as both a development tool and a hazard. This interdependence has implications for developing comprehensive conceptualizations of wildfire risk.

A comprehensive conceptualization of wildfire risk requires two inputs: The first relates to forest ecology. The second is concerned with understanding the determinants of the ability of people, communities, and societies to be able to coexist harmoniously with all facets of the amenity and hazardous aspects of the forest environment. That is, it is important to bring a social—ecological perspective to bear on the task of understanding and managing wildfire risk. It is thus becoming increasingly important to consider the contributions to risk emanating from environmental, ecological, and social domains as well as the interdependencies between them (Buergelt and Paton, 2014; Paton, 2006; Paton et al., 2006). The contents of this volume focus primarily on the social—ecological contributions to wildfire as a hazard.

Before proceeding to introduce the content and its social—ecological implications, a brief note on nomenclature is warranted. The term wildfire will be used predominantly in this volume. However, while this term predominates in North America, the terms bushfire and forest fire are used in Australia and Europe, respectively. The terms bushfire and forest fire are used where appropriate. Whatever the term used to describe the phenomenon, the risk wildfire poses is increasing and will continue to do so for the foreseeable future.

1.2 CHANGES IN THE WILDFIRE HAZARD SCAPE

It is undeniable that wildfire presents a growing risk to many countries. The contributions to this volume illustrate this with respect to what is happening in

North and South America, Australasia, India, Europe, and Russia. In addition to this increased risk emanating from the ubiquitous influence of climate change on the fundamental environmental and ecological sources of wildfire risk (e.g., hotter temperatures over longer periods of time), social change (e.g., population growth in the wildland—urban interface (WUI)) is making a contemporaneous contribution to increasing the scale, consequences, and duration of the ecological, economic, and social consequences that will continue to arise from wildfire events. These factors, in turn, interact to conspire to create ever-more complex disaster risk reduction (DRR), disaster recovery, and restoration challenges in countries susceptible to experiencing wildfire hazards.

While wildfire risk management issues are most immediately apparent in North America, Southern Europe, and Australia, wildfires are increasingly making their presence felt in many other countries. Societies and communities susceptible to experiencing wildfire hazards can expect not only more frequent large-scale, damaging wildfire hazard events but also fires of greater intensity and duration. The potential scale of this emerging threat is reflected in the increasing occurrence of megafires (Adams, 2013). Megafires deserve special attention in both fire management and postfire restoration (see Chapters 5 and 12). The potential occurrence of megafires highlights the potential for wildfire to create global impacts through the emissions of greenhouse gases and particles to the atmosphere.

Climate change will influence the patterns of wildfire risk and their distribution (e.g., Nicholls and Lucas, 2007). Countries where this hazard was infrequent can expect more, and some countries will find themselves having to add wildfire to their national hazard scapes for the first time. Even if not directly affected, wildfire hazard consequences can readily transcend national borders and create impacts on a global scale.

Locally, wildfire consequences affect air quality, ecosystems and landscapes, and the built environment. Secondary consequences can be created by the impact of fire hazards on soil and water quality. These, in turn, can affect food and water insecurities. Globally, wildfires create large-scale problems through their contribution to atmospheric emissions of greenhouse gases and particles. These can create direct damage to vegetation and fauna, direct and indirect impacts on soils through heat release and ash deposition, and contribute to postfire environmental degradation (see Chapter 12). Global impacts are increasingly evident from the ways in which wildfire smoke (e.g., from Indonesian fires—see Chapter 6) transcends national borders and can affect whole regions.

The contents of this book review the findings of substantive research programs on the social—ecological dimensions of wildfire risk. The contributors bring their considerable expertise and experience of researching wildfire to discuss the personal, social, societal, environmental, and ecological factors that need to be accommodated in wildfire risk management. By drawing on empirical work, the contributions to this volume provides the kind of robust evidence base required by fire, land use, and other government agencies responsible for wildfire mitigation and risk management to design effective mitigation, risk communication, community outreach, threat assessment, warning, environmental management, and postfire restoration programs in communities that find themselves, by accident or design, having to live and prosper in environments in which the forest environments they rely on for resources, livelihoods, lifestyle, and recreation periodically become hazardous. The global implications of helping countries realize this goal calls for international contributions. This process commences in the Americas.

1.3 THE AMERICAS

1.3.1 The United States of America

McCaffrey, Toman, Stidham, and Shindler (Chapter 2) introduce the complex web of social and community contributions to wildfire risk and discuss their application to strategies that can be used to conceptualize social–ecological approaches to wildfire risk management. In doing so, McCaffrey and colleagues highlights how wildfire risk derives from and can only be managed by accommodating the dynamic interdependencies that arise between people, societies, and the environments in which exist (and which change over time). Accommodating these interdependencies creates the essential foundation for facilitating the development of fire-adapted communities that accept their risk and their responsibility for managing their risk through actively engaging residents, community leaders, and a range of governmental and nongovernmental organizations in risk management. To this, McCaffrey et al. illustrate the benefits that can arise from analyzing wildfire risk, at least in part, through a historical lens.

McCaffrey and colleagues discuss how Native Americans were active resources managers. While they used fire to, for example, stimulate the production of desired plant species, decrease disease and pests, and facilitate game hunting, the arrival of early-Euro-American settlers saw a progressive shift from using fire to manage resources to starting to organize ways to suppress fire as permanent settlements were established. The latter paved the way for the societal development of policies and practices that have, over time, shifted from highly interventionist fire control polices to more comprehensive fire management (e.g., restoration of fire-adapted ecosystems, reduction of wildland fuels, and provision of economic assistance to rural communities) practices.

The development of such policies introduces the complex governance, multijurisdictional and multiagency coordination, and practices that arise from turning this policy into practice. McCaffrey and colleagues discuss how strategies such as the national Cohesive Wildfire Management Strategy act to facilitate the attainment of collaborative practices between diverse stake-holders and offer solutions to fire management problem in three specific areas: Restore and Maintain Landscapes, Fire-Adapted Communities, and Response to Fire.

Chapter 2 sets the scene for understanding how wildfire risk will increase in the foreseeable future as a result of ecological changes created by fire suppression, climate change, growing number of people moving into fringe of urban areas or in more rural wildland areas, and fragmented property ownership. Effective fire management will require transcending numerous boundaries of land ownership and understanding and accommodating the perspectives of diverse stakeholders with regard to prefire social dynamics (e.g., risk acceptance, support for mitigation, demographic/gender issues, collaborative risk management, and trust in agencies) and during and postfire social dynamics (e.g., evacuation vs "Shelter in Place" model, mass and social media, community engagement, and ecological restoration). Collectively, this discussion highlights the need for using community engagement and empowerment principles to integrate the perspectives of multiple stakeholders into a risk management program.

1.3.2 Canada

In chapter 3, coverage of North American wildfire issues continues and in-troduces a Canadian perspective. McGee, McFarland, and Tymstra discuss how wildfire in Canada occurs in forest landscapes that are integral to Canadian culture and that are an important resource for the Canadian economy and society. While Canada has so far been spared regular major losses from wildfire events (e.g., as a result of effective wildfire suppression and low population densities in areas with high wildfire activity), population growth, and industrial development that is encroaching on forested areas is acting to increase wildfire risk in the region. McGee and colleagues how development pressures will be compounded by the effects of climate change. Collectively, these pressures are acting to increase, the future risk to communities and infrastructure as a consequence. This, in conjunction with the growing awareness of wildfire suppression costs, has resulted in Canada putting more emphasis on mitigating the risk before wildfires occur. As with their southern neighbor (see Chapter 2), enacting this approach will require improved collaboration and cooperation among stakeholders across the country. They reiterate (see Chapter 2) the challenges and benefits that can arise from un-derstanding how Aboriginal peoples in Canada (First Nations, Métis, and Inuit) have a history of using fire dating back some 8500 years to learn about and utilize indigenous knowledge in the development of wildfire DRR.

McGee and colleagues provide further insights into the importance of accommodating multijurisdictional and multiorganizational issues in risk management planning. An important organizational issue they identify reflects the reliance of many rural community fire departments on volunteer fire fighters and how this adds an additional dimension to interagency dynamics (e.g., through a need for mutual aid agreements with neighboring jurisdictions and provincial and territorial wildfire agencies and local industries). This work also introduces a need for community engagement strategies to expand to encompass the development of a sustainable volunteer resource. The importance of addressing this issue is heightened by the fact that volunteers can make additional contributions to promoting community readiness.

McGee et al. discuss how recognition that wildfire is essential to healthy ecosystems suggest a new approach is needed for wildfire management in Canada. McGee and colleagues discuss this in terms of the need for better shared vision for wildfire management among provincial, territorial, and federal governments and discuss how emergent organizations, such as the Partners-in-Protection Association, are making complementary contributions to how provincial, territorial, and municipal agencies can facilitate community education, encourage community-based initiatives, and the engagement of diverse stakeholders in risk mitigation.

1.3.3 South America—Chile

Chapter 4 sees the focus turn to South America, to Chile. There too, an upward trend in wildfires is occurring as a consequence of the increasingly intensive use of renewable natural resources, whether for production or as a source of recreation and entertainment. The presence of extensive WUI areas is a constant and increasing problem for fire spread risk and danger.

This provides the context in which Miguel Castillo Soto, Guillermo Julio-Alvear, and Roberto Garfias Salinas discuss the development of a Chilean Danger Evaluation System. The Wildfire Forecast System, based on past scientific records and data from the previous 30 years, KITRAL, is one of the main technological advances in fire management developed in Chile. It comprises five modules covering fire detection, allocation of fire fighting resources, calculation of territorial priorities for fire protection, risk forecast model, and wildfire simulator.

1.4 EUROPE

Chapter 5 provides a discussion across the Atlantic to see that, as with their American counterparts, all European states face a growing risk from increasing population density, creeping urban sprawl, changes in land-use patterns that conflict between societal and ecological protection, and changing climatic and weather conditions. Likewise, the European states face the challenge of reconciling social, economic development, environmental concerns, and with living with forest fires in a sustainable and dynamic equilibria with their risk management frameworks. Tedim, Xanthopoulos and Leone argue that this is best done by understanding historical contributions to the social construction of wildfire risk, and doing so in ways that capitalize on Traditional Ecologic Knowledge (Ribet, 2002).

While, and in common with other countries grappling with wildfire risk, European countries face a growing risk from Wildland—Urban Interface (WUI) expansion, Tedim and colleagues suggest caution in assuming that the contribution to risk from this source in Mediterranean Europe is the same as in, for example, Australia, Canada, and the United States. The distinction, they suggest, is mainly due to the differences in built environmental characteristics. This observation highlights the need to the systematic analysis of the complex web of interdependent factors that contribute to risk and not assuming that the same factors make the same contribution to risk despite the existence of superficial similarities.

Tedim and colleagues introduce a sense of urgency into the task of wildfire risk management as a result of the growth in the occurrence of Mega Fires. These very large fires have become a part of the European reality since the early 1990s and, in the context of the implications of climate change, will become more common in the future. The fact that these megafires overwhelm the capability and endurance of available fire fighting resources makes a compelling argument for risk reduction and mitigation efforts.

The adoption of a whole-of-Europe perspective in Tedim et al.'s chapter reiterates the contribution that multijurisdictional arrangements make to risk management. Fire management is complicated by the variety of fire management organizations in various European countries and by diversity of forest ownership (e.g., federal, state, municipalities, monasteries, and private). Tedim and colleagues discuss how mitigation and fire defense needs are often inconsistent with organizational structure, political—administrative decentralization, and organizational culture. This reiterates the need to include the development of multijurisdictional and agency relationships in future research and intervention agenda. They conclude by discussing how the broader implications of progressive environmental and climate change necessitates attention being directed to, for instance, fire prevention through landscape planning and connectivity, infrastructure design, training, and the development of international collaboration on forest fire prevention and fire fighting for crossborder fire events. These issues are also highly pertinent to several countries that lie south of the equator.

1.5 AUSTRALASIA

1.5.1 Indonesia

Perspectives on the social—ecological aspects on Australasian wildfire risk and its management are introduced by Sagala, Sitinjak, and Yamin in Chapter 6. Their discussion is framed in terms of how forests are a major component of the Indonesian economy and how interaction between people and forests, while driven predominantly by economic activities, is creating sources of wildfire risk. For example, forest clearing practices linked to commercial activity (e.g., palm oil production) has increased wildfire risk as a result of slashing techniques leaving behind increasing fuel loads and creating drier soil conditions that are more susceptible to fire, especially in peat soils. Indonesian wildfires also create serious secondary hazards, such as smoke (both within Indonesia and in neighboring countries), with this contributing to both health problems and environmental degradation.

Sagala et al. echo earlier comments and discuss how Indonesian approaches to risk management that have focused on suppression, rather than tackling the underlying causes of wildfires, have created divergent views about the effectiveness of community fire management. They argue for a need to switch to community-based practices that integrate indigenous knowledge, conservation values, and sustainable livelihoods through research that builds an understanding of (1) ecology and wildfire behavior; (2) the needs and behaviors of community members; and (3) relationships between fire and the community. They conclude by discussing how an Indonesian Ministry of Forestry initiative, the "Hutan Taman Rakyat" scheme (which translates as community garden forest) represents an approach to developing effective fire management by ensuring that local communities living nearby or in forests are the ones who manage and take charge of forest management.

1.5.2 Australia

Just to the south of Indonesia lies Australia, a country for which wildfire is a natural and inevitable part of its ecosystem. Buergelt and Smith discuss how Australia's topographic characteristics, its vegetation, and its weather make it one of the most fire-prone areas in the world, a view that is reflected in it facing wildfire-related costs of some $1.6 billion per year. They follow the pattern in the North American and European discussions and illustrate how traditional peoples' sustainable use of fire and ability to live harmoniously with fire were significantly affected by European settlers ignoring local knowledge, applying fire management strategies based on European climate and agriculture. By clearing native endemic vegetation and restructuring fuels, European settlers unintentionally increased the risk of wildfires due to a lack of local knowledge of the ecosystems.

Buergelt and Smith proceed to discuss how one of the most complex and divisive issues in Australian WUI communities derives from stakeholders holding diverse beliefs regarding the management of vegetation to reduce the fire hazard and the management of the vegetation for conservation. Buergelt and Smith discuss how getting the balance in between competing beliefs is a crucial challenge and one whose importance and urgency are being fueled by the growing contribution of climate change to wildfire risk.

Buergelt and Smith add weight to the need to consider the multiagency and multijurisdictional aspects of risk management. Their argument in this regard derives from the observation that collaboration and coordination among and between diverse stakeholders (federal and state government departments, associations and peak bodies, NGOs and service providers, local governments, social and natural hazard researchers, public and private education providers, media and businesses) across all scales and across states/territories is not as effective as it could be. A clear need for multiagency integration and community engagement is thus indicated.

Buergelt and Smith argue that to resolve the "wicked problem" that is wildfire risk, there is a need for new overarching strategies to facilitate collaboration and capacity building. They discuss how new mind-sets are required to facilitate collaboration (e.g., sociocracy) and promote comprehensive risk management solutions to complex problems. They close by outlining how novel learning and teaching approaches and communication technologies can be used to facilitate the education and training of professionals, volunteers, and communities in ways that support capacity building. While Australia has a long history of fire, the next chapter introduces a country in which fire is an emerging hazard, that is, Taiwan.

1.5.3 Taiwan

Chen and Chen open by discussing how Taiwan's island climate, and one characterized by high humidity, means that forest fires do not easily occur. However, it is becoming a significant hazard Consequently, because of the high resource value attached to Taiwanese forests, a need for effective management and postfire restoration plans and strategies has been recognized.

Chen and Chen discuss how, as a consequence of Taiwanese wildfire risk being situated in steep and remote terrain, current management efforts are focusing on the application of geographic information system (GIS) and Aerial Telemetry Information in Forest Fire Predictions. They continue by discussing how the Taiwan Forestry Bureau has applied GIS to "develop decision support systems" in forest fire management situations. The novelty of wildfire hazards in Taiwan has also prompted researchers and practitioners to see what could be learned from elsewhere. In doing so, Chen and Chen highlight how, for example, international inconsistencies in Forest Fire Danger Rating Systems limited opportunities for international crossfertilization of wildfire risk and danger rating research and its application. From a social perspective, wildfire management in Taiwan requires taking diverse cultural and spiritual factors into consideration. This significant issue is discussed further in the next chapter.

1.6 INDIA

In India, wildfires make a relatively minor, but growing, contribution to the hazard scape. As in other countries, India's wildfire risk is being influenced by increases in population-driven changes in land use and by migration of people into formerly untouched forest areas. Hazardous fires in sensitive mountain ecosystems then create secondary hazards, such as landslides, mudslides, erosion, increased water run-off, flash floods, soil depletion, loss of cultural heritage sites, and health problems associated with smoke.

Schmerbeck and Kraus discuss how the source of wildfire risk derives from fire being an essential tool for landscape management, with India's landscape as it appears today being largely a product of extensive historical fire use practices. The landscapes have been shaped by a long history of human-influenced fire regimes and coevolved with fire and many of the plants have adapted to fire. It continues to be important for local livelihoods and the production of important domestic and commercial goods.

Schmerbeck and Kraus echo earlier thoughts of historical contributions to risk by discussing how it was that the introduction of European fire practices by the British Raj shifted the balance from sustainable use to practices of fire exclusion and suppression. These historical practices contributed to increasing risk. As a result of recent recognition that colonial fire practices were counterproductive saw attempts, which were never 100 percent effective, to reintroduce fire as a forest management practice.

The Indian case study reiterates the complexity inherent in social–ecological conceptualizations of wildfire risk and its management. Schmerbeck and Kraus discuss how in contemporary India, wildfire risk and thus its management reflect interactions between several factors. These include cultural and spiritual factors, economic activity, people's fuel needs, the need for public safety in forests and nearby communities, and pastoral and agricultural issues. Integrating these issues into a coherent strategy, they suggest, requires considering the needs that prevail in local settings, people's fire-use motivation, and how the ecosystem responds to fire regimes. In India, a significant challenge to wildfire risk management derives from a high population density. In the next chapter, the tables are turned, and it is seen that low population density is one of the major culprits.

1.7 RUSSIA

Ponomarev, Ivanov, and Korshunov open by discussing how Russia's size and the significance of wildfire make unique contributions to its hazard scape. The fact that this can arise in forest, forest-steppe, and steppe zone gives Russia a unique wildfire risk profile. The need for effective management in Russia is framed in terms of how climate change will increase extreme wildfire-related emissions (CO_2) in Siberia. Forests in the northern territories of Siberia have a significant economic and ecological importance as one of the major components of the global carbon budget. Ponomarev and colleagues argue that managing this risk is complicated by low population density and extreme access difficulties. This results in a high level of burning and a significant number of extreme large-scale wildfires annually, with the associated escalating pattern of risk increasing the need for risk management and monitoring. The vast nature of the geographical area to be managed calls for strategies that utilize, for example, on-ground and air monitoring and the use of satellite monitoring to expand the area capable of being monitored. The effectiveness of these activities can be facilitated by improved planning and warnings. This issue is the subject of the next chapter.

1.8 WILDFIRE DANGER RATING AND WARNINGS

The management of the social—ecological complexity and geographical scale of factors contributing to wildfire risk can be facilitated by having a means rating fire danger rating and by being able to provide warnings. De Groot discusses how Danger ratings and warnings form the foundation for decision making in many contemporary fire management programs. De Groot discusses how Fire danger information is used across a broad spectrum of fire management decision making, including daily operations, seasonal strategic planning, and long-term fire and land management planning under future climate change. De Groot continues by discussing warning systems, their diversity, and how they differ with regard to the function they perform and the stakeholder needs they serve. At national, regional, and global scales, this encompasses activities that occur over long time frames (days to weeks in advance) to make large-scale decisions such as planning national or international mobilization and distribution of suppression resources.

The scale of current wildfire risk and the fact that it is only likely to increase in the future make fire an inevitability in many parts of the world. The importance of forest environments makes it important to understand how environmental restoration can be understood and facilitated. This topic is tackled in the final chapter.

1.9 RESTORATION

A theme running through the preceding chapters reflects the strong interdependencies that exist between people and communities and the forest sources of wildfire hazards. This means that the loss of forests and forest resources creates losses on several socioecological levels. Thus, the concept of wildfire risk extends beyond the prefire context to include postfire issues, with postfire restoration thus making particularly significant contributions to social and ecological recoveries. In the final chapter, Vallejo and Alloza tackle this issue and discuss how restoring resilient landscapes (e.g., soil productivity, recuperation of prefire vegetation) will reduce damages and suppression costs in the long term. Vallejo and Alloza continue by arguing that postfire restoration projects should be carefully prioritized, and scientifically and technically justified, with postfire restoration assessment being based on the understanding of how fire regime is affecting ecosystem fire resilience. They conclude by discussing how the growing prospect of megafires adds to the urgency of prioritizing this issue in wildfire risk management planning.

1.10 DEVELOPING A SOCIAL–ECOLOGICAL PERSPECTIVE

Collectively, the contributions to this volume illustrate how and why wildfire risk and its management must be conceptualized as a social–ecological phenomenon. The chapters illustrate how the ecological features of forests can interact with dynamic environmental conditions (e.g., interaction between vegetation, topography, and weather) to create hazards whose impact and implications are a function of how the specific mix of historical, psychological, social, cultural, spiritual, economic, political, and organizational characteristics in at-risk communities create complex risk management contexts. The nature and extent of the risk communities experience is further affected by factors such as understanding how risk is distributed (e.g., using danger ratings) and how communities can be forewarned (e.g., via warnings) to ensure that they can enact mitigation and response strategies in timely and effective ways. Comprehensive risk assessment additionally requires appreciation of how long and to what extent environmentally, social, and economically significant forest environments can be restored. These are summarized in Figure 1.1.

The placement of people at the center of Figure 1.1 defines how risk emanates from the choices that people make within their social–ecological environment. The fact that several of these factors are amenable to intervention and/or provide a cultural foundation for effective response opens up several possibilities for DRR (Paton and Tedim, 2012; Tedim and Paton, 2012). It is important to note the need to consider an additional temporal dimension when considering the conceptual model described in Figure 1.1. The nature of the content of each segment and how they relate to each other will change over time. Wildfire risk management is an iterative process that occurs in a dynamic social–ecological context.

Understanding how these features contribute to both risk and how risk can be managed play pivotal roles in how wildfire risk is managed. In addition to accommodating contemporary issues, discussion of indigenous practices in several chapters introduced the potential lessons that can be learned by seeing ourselves as part of the natural ecosystem through creating a greater sense of

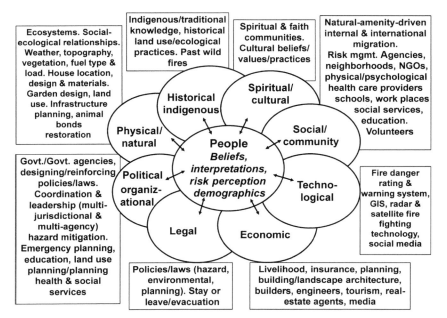

FIGURE 1.1   Summary of social ecological contributions to wildfire risk.

connection with and shared responsibility for valuing and respecting fire and developing risk management philosophies and practices that facilitate coexisting with the forest sources of risk.

Further complexity is introduced by the need to define and manage wildfire risk in terms of the influence of, and interaction between, the myriad social and ecological characteristics that influence risk (Figure 1.1). Accommodating these factors and the interactions between them make the risk management process a multidisciplinary and multifaceted one in which the goals, needs, and capabilities of diverse stakeholders are accommodated in research and in planning, intervention design and delivery, as well as in postfire restorative and recovery interventions. The pursuit of this more comprehensive approach wildfire risk management, particularly with regard to recognizing our frequent failure to learn historical lessons, will benefit from heeding to Einstein's warning that "we can't solve problems by using the same kind of thinking we used when we created them"; effective wildfire risk management requires all stakeholders to learn to "think outside the square." Doing so will becoming increasingly important if we are to safeguard people and the environment from the growing wildfire risk emanating from climate change processes. The contents of this volume provide a sound foundation for doing so. How to achieve this is introduced in the next chapter.

REFERENCES

Adams, M.A., 2013. Mega-fires, tipping points and ecosystem services: managing forests and woodlands in an uncertain future. For. Ecol. Manage. 294, 250–261.

Buergelt, P.T., Paton, D., 2014. An Ecological Risk Management and Capacity Building Model. Hum. Ecol. 42, 591–603.

Clayton, S., Opotow, S., 2003. Identity and the Natural Environment: The Psychological Significance of Nature. MIT Press, Cambridge.

Nicholls, N., Lucas, C., 2007. Interannual variations of area burnt in Tasmanian bushfires: relationships with climate and predictability. Int. J. Wildland Fire 16 (5), 540–546.

Paton, D., 2006. Disaster resilience: building capacity to co-exist with natural hazards and their consequences. In: Paton, D., Johnston, D. (Eds.), Disaster Resilience: An Integrated Approach. Charles C. Thomas, Springfield, Ill, pp. 3–10.

Paton, D., Kelly, G., Doherty, M., 2006. Exploring the complexity of social and ecological resilience to hazards. In: Paton, D., Johnston, D. (Eds.), Disaster Resilience: An Integrated Approach. Charles C. Thomas, Springfield, Ill, pp. 190–209.

Paton, D., Tedim, F., 2012. Wildfire and community: facilitating preparedness and resilience. Charles C. Thomas, Springfield, Ill.

Ribet, N., 2002. La maîtrise du feu un travail "en creux" pour façonner les paysages. In: Woronoff, D. (Ed.), Travail et paysages, Paris, Éditions du CTHS, Actes du 127e Congrès du CTHS, Le travail et les hommes, Nancy 15–20, April 2002, pp. 167–198.

Tedim, F., Paton, D., 2012. A Dimensao Humana dos Incendios Florestais. Estratégias Criativas, Porto, Portugal.

Chapter 2

Social Science Findings in the United States

Sarah McCaffrey
Northern Research Station, USDA Forest Service, Evanston, IL, USA

Eric Toman
School of Environment and Natural Resources, The Ohio State University, Columbus, OH, USA

Melanie Stidham
School of Environment and Natural Resources, The Ohio State University, Columbus, OH, USA

Bruce Shindler
Department of Forest Ecosystems and Society, Oregon State University, Corvallis, OR, USA

ABSTRACT

The rising number of acres burned annually and growing number of people living in or adjacent to fire-prone areas in the United States make wildfire management an increasingly complex and challenging problem. Given the prominence of social issues in shaping the current challenges and determining paths forward, it will be important to have an accurate understanding of social dynamics. After providing a brief contextual background of fire management in the United States, this chapter focuses on a review of the key findings from social science research related to how the public views fire management in the United States. Primary topics discussed are public acceptance of fuels treatments on public lands, homeowner mitigation activities, and social dynamics during and after a fire. The goal of the chapter is to (1) provide fire managers and other interested stakeholders with an accurate understanding of what shapes public response to fire management before, during, and after fires; (2) provide a context for future research; and (3) inform future efforts to foster fire-adapted communities where people are aware of the fire risk and have taken appropriate action to reduce that risk and increase resilience to wildfire.

2.1 INTRODUCTION

Wildland fire management in the United States is an increasingly complex and challenging problem. With a rising number of acres burned annually and a growing number of people living in or adjacent to fire-prone areas, much is at

Wildfire Hazards, Risks, and Disasters. http://dx.doi.org/10.1016/B978-0-12-410434-1.00002-6

stake. Creating fire-adapted communities, where there is an awareness of the wildfire risk and actions have been taken to mitigate that risk and increase resilience, will require the active participation of residents, community leaders, and a range of governmental and nongovernmental organizations. Given the prominence of social issues in shaping the current challenges and in determining paths forward, it will be important to have an accurate under-standing of social dynamics.

This chapter will first provide a brief contextual background of fire management in the United States. It will then provide a review of the key findings from the social science literature related to fire adapted commu-nities, including acceptance of mitigation action on public lands, homeowner mitigation activities, and social dynamics during and after a fire. Our intent is to provide fire managers and other interested stakeholders with an accurate understanding of what shapes public response to fire management throughout the fire management cycle, inform efforts to foster fire adapted communities, and provide a context for future research.

2.1.1 Historical Context

Underlying the current fire management challenge in the United States is a fire policy that for most of the twentieth century has focused on fire control or full suppression. Prior to this focus, studies have shown that Native Americans were active resource managers, who used fire for a variety of reasons, including to stimulate production of desired plant species, decrease disease and pests, and facilitate game hunting (Huntsinger and McCaffrey, 1995; Lewis, 1993). Early-Euro-American settlers also extensively used fire to manage resources, but as permanent settlements were established, they also began to organize to suppress fires that threatened private resources (Pyne, 1997).

Fire suppression first took hold as a formal policy at the turn of the twentieth century, in congruence with the advent of the Progressive Era. Growing public concern over mismanagement and potential scarcity of natural resources led to calls for greater government oversight and management of the nation's natural resources to maximize present and future use. This movement led to the creation of the Forest Reserve Acts of 1891 and 1897, which withdrew large tracts of timber from settlement. In 1897, the Bureau of Forestry was created to manage the reserved areas, and, in 1905, the Bureau was transferred to the Department of Agriculture and became the United States Forest Service (Hays, 1959). From its inception, the agency was staffed by professional foresters whose training was based on methods imported from Europe where forests had been actively managed for centuries and fire suppression was an integral part of management practices (Behan, 1975; Pyne, 1997).

In 1910, a series of massive fires in the Northern Rockies consolidated the emphasis on fire suppression. Over subsequent decades, fire control ef-forts expanded with full fire suppression finally achieved with the arrival of

World War II. The war meant that forest products were increasingly valuable and also raised fears of Japanese incendiary bombs setting fire to the West Coast. Putting out all fires now became a patriotic duty (slogans of the time included "Careless Matches Aid the Axis" and "Another Enemy to Conquer, Forest Fires"), and advertising agencies were enlisted to develop fire prevention ad campaigns. This culminated in the creation of Smokey Bear in 1945 (Pyne, 1997).

Changes in wildfire policy began in the 1970s with the growing recognition of the technological limits to full suppression and of the important ecological role that fire plays in many forest ecosystems (Pyne, 1997; Davis, 2001). In many ecosystems, removing fire had affected the type, density, diversity, and pattern of vegetation, generally in a way that added to the fire hazard, particularly in terms of fuel load buildup. The policy shift was also due in part to an increase in the fire management responsibilities of federal agencies other than the Forest Service, which introduced other agency viewpoints into the mix. The Bureau of Land Management (BLM) entered into fire control in a significant manner in the 1950s when it began to try to control fires on its Alaska lands, which made up half of all federal lands (Pyne, 1997). The National Park Service (NPS) began experimenting with prescribed burning in the Everglades in the 1950s and in 1968 began to introduce let-burn and prescribed fire programs into its parks with the goal of restoring the ecological role of fire (USDI/USDA 1995).

In 1977, the Forest Service formally changed its fire policy from fire control to fire management and prescription fires, planned or natural, became a formal part of fire management policy (Pyne, 1997). In 1988, large fires in the Yellowstone National Park focused attention of the media, political officials, and the general public on wildfire-management strategies and prompted reviews of fire policies (Davis, 2006). A 1995 revision of existing federal fire policy recognized the role of fire in ecological systems and called for implementation of fuel reduction programs to reduce the likelihood of catastrophic fire events (Stephens and Ruth, 2005). A series of large wildfires in 2000 prompted additional policy revisions (Moseley, 2007).

Under the suppression-centric approach, the wildfire-management authority rested almost exclusively with federal resource management agencies; however, more recent efforts have emphasized greater intergovernmental coordination in prefire preparations and during-fire management (Davis, 2001). As more nonfederal lands have been impacted by wildfires, state agencies and local fire departments have become more involved in wildfire-management issues. Although the Forest Service remains the largest wildland fire-fighting agency, today wildfire management in the United States must take into account a complicated mix of five federal agencies with fire management responsibilities (USFS, BLM, NPS, Fish and Wildlife Service, and the Bureau of Indian Affairs), state forestry agencies, and a vast network of independently operated local (county, municipal, and volunteer) fire departments.

Recent policies including the National Fire Plan (2000); the Western Governor's Association 10-Year Comprehensive Plan (2001); and the Healthy Forest Restoration Act (2003) reflect the shift away from a policy of complete fire suppression to one that includes a broader set of goals including restoration of fire-adapted ecosystems, reduction of wildland fuels, and providing economic assistance to rural communities (Gorte, 2003; Steelman et al., 2004). Federal and state agencies are involved in determining the resources available to mitigate risk at the local level; the federal government largely sets policy direction and provides financial resources, while state governments make organization and programmatic decisions about how to allocate those resources to mitigate fire risk (Steelman et al., 2004). Despite the development of this shift toward a broader set of policy goals than just suppression than evaluations of fire management practices suggest that, in practice, fire suppression and hazardous fuels reduction receive the most attention and resources, sometimes at the expense of restoration and community assistance (Gorte, 2003; Steelman et al., 2004; Jensen, 2006; Steelman and Burke, 2007).

More recently, the 2009 Federal Land Assistance Management and Enhancement Act mandated that federal agencies work with stakeholders to develop a national Cohesive Wildfire-Management Strategy. The Strategy uses a collaborative process to bring together stakeholders from all levels—local to national and governmental to individual homeowners—to develop local-, regional-, and national-level solutions to the fire management problem in three specific areas: (1) restore and maintain landscapes; (2) fire-adapted communities; and (3) response to fire. While the final phase of development has just recently been completed, the Strategy is expected to influence fire management and related objectives and allocation of resources on the ground.

2.2 REVIEW OF RELEVANT RESEARCH FINDINGS

In many places in the United States, the ecological changes created by fire suppression have contributed significantly to the growing severity of fires. Ongoing environmental change, including warmer spring and summer temperatures, increased drought, earlier spring snowmelt, and longer fire seasons have further contributed to changed fire activity (Westerling et al., 2006). These two elements, in and of themselves, would create a highly challenging wildfire-management problem. However, management is further complicated by the increasing impact of diverse social concerns particularly, the growing number of people moving into high fire-hazard areas, either on the fringe of urban areas or in more rural wildland areas—regions that are often referred to as the wildland urban interface (WUI). The presence of more people in fire-prone areas presents significant complications in fire fighting and in determining ways to mitigate the hazard. At the most basic level, more humans and more structures create more values at risk from a wildfire. Fragmented property ownership also creates challenges for consistent management among diverse owners. Thus, effective fire management will require transcending numerous boundaries of land ownership and an understanding of the perspectives of diverse stakeholders. However, as a relatively new focus in fire management, there are many assumptions about how social elements influence current fire management dynamics. Understanding the accuracy of these assumptions will be important in developing policies that most effectively decrease future negative outcomes from wildfire.

The remainder of this chapter provides an overview of recent scientific findings in relation to understanding public perspectives of wildfire management. Although social scientists have conducted research on the human dimensions of wildland fire management for >40 years, the vast majority of this work has been conducted since 1998. The main focus of research to date has looked primarily at mitigation activities on either public or private land before a fire occurs. A smaller but growing body of work has begun to examine social dynamics during and after a wildfire. This chapter summarizes the findings in these areas from more extensive documents developed from a review of >200 research articles for a project that was funded by the Joint Fire Science Program to take stock of the key findings from social science research over the past decade. For ease of reading, we have provided minimal citations in this chapter; specific information about the range of studies being referenced can be found in the longer documents (Toman et al., 2013; McCaffrey and Olsen, 2012).

A consideration in interpreting findings is that, with only three exceptions, research participants in the studies lived in or owned homes in the WUI and thus may be more aware of wildland fire than perhaps the general public. However, findings across studies suggest that such geographic distinctions may not be that meaningful in understanding differences in response to wildfire. How much effect geographic and sociodemographic differences may have on public response is discussed more fully at the end of the chapter.

2.3 PREFIRE SOCIAL DYNAMICS

Overall, studies find that residents of fire-prone areas generally recognize the fire risk and often have a sophisticated understanding of the ecological role of fire. Many studies found that participants' comments indicated a good understanding of fire behavior and fire ecology and of the various factors that contribute to fire risk. For example, Vining and Merrick (2008) found that Minnesota study participants understood the complex nature and tradeoffs of different fire-management practices and understood "that fire-management techniques have just as many (or perhaps more) ecological benefits as negative ecological consequences." In qualitative studies, understanding of the role of fire in the environment is referenced primarily in relation to three related topics: (1) awareness of the risks of living in a natural landscape; (2) perceptions that the current forest is unhealthy from too many trees and/or a buildup of fuel; and (3) discussions of overall forest management and the need to reintroduce fire, whether via prescribed fire or allowing some naturally ignited fires to burn. A number of studies have found that forest health is generally a parallel and sometimes more dominant consideration than reducing fire risk in shaping an individual's response to management actions.

2.3.1 Fuel Management on Public Lands

A significant number of studies have examined public acceptance of fuel management practices on public lands, primarily in relation to prescribed fire and mechanized thinning practices. Studies in a variety of locations have found high levels of acceptance (>80 percent in many at-risk communities) for some use of prescribed fire and/or of mechanized thinning treatments. A series of studies (Shindler and Toman, 2003; Brunson and Shindler, 2004; Shindler et al., 2009, 2011) that found overall acceptance levels of >80 percent distinguished between unqualified acceptance (legitimate tool, use anywhere) and qualified acceptance (use in carefully selected areas). For prescribed fire, roughly equal proportions of respondents chose unqualified and qualified acceptance, while for thinning, a greater proportion chose unqualified acceptance (50 percent vs 30 percent).

Although the majority of studies found high acceptance levels for both treatments, a few studies have found more measured levels of support; some have found lower acceptance levels for prescribed burning and others for thinning. For instance, a nationwide survey of the general public asked whether participants agreed or disagreed with manager use of prescribed fire and mechanical vegetation removal as part of a wildfire-management program. Nearly all participants agreed with the use of prescribed fire (91 percent), while fewer, though still a majority (58 percent), agreed with the use of mechanical vegetation removal (Bowker et al., 2008). Conversely, Toman et al. (2011) found high levels of support for thinning (83 percent) and lower levels of support for prescribed fire (66 percent).

Substantially fewer studies consider public acceptability of other fuel reduction methods. The largest body of findings indicates that taking "no action" is consistently the least preferred choice and that acceptance levels for herbicide use are fairly low, with large proportions of respondents finding herbicide use unacceptable. Only a few studies included livestock grazing as a treatment. These found that it is a generally acceptable practice with roughly 80 percent indicating partial or full acceptance. For both grazing and herbicides, rural respondents tended to be more supportive of the practice than respondents from urban areas.

Limited research has examined public acceptance of managing unplanned ignitions to achieve resource benefits. One-third to one-half of visitors to three National Forests in California, Colorado, and Washington agreed with allowing naturally ignited fires to burn when the fire was expected to result in minimal impacts to human communities or the forested ecosystems (Kneeshaw et al.,

2004). Similarly, a survey of California residents found that 60 percent agreed with allowing some fires to burn as long as residences were protected (Winter, 2002). Research focused on fire managers has also identified a number of factors that may limit adoption of this practice including psychological factors (e.g., attitudes of fire managers favor suppression, risk of personal liability) and other policy-related factors (e.g., extensive planning requirements, need for specialized personnel, inability to qualify for emergency stabilization funds should something go wrong, air quality regulations).

While not universal across all studies, some findings suggest that treatment acceptance can differ depending on the specific location of treatment implementation. The few studies that examined acceptability of letting naturally ignited fires burn have found higher levels of acceptance for the practice in remote areas removed from private lands. Similarly, a number of studies have found a preference for the use of mechanical thinning in more urbanized areas and for use of prescribed fire in less populated areas. However, one study found similar levels of acceptance for use of prescribed fire both in remote areas and around neighborhoods (Toman et al., 2011). One study also found that land ownership or designation can play a role in acceptance with respondents indicating a preference for the use of prescribed fire on NPS lands versus a slight preference for the use of mechanical harvest (preferably in conjunction with prescribed fire) on Forest Service and private lands (McCaffrey et al., 2008).

2.3.1.1 Concerns with Potential Treatment Impacts

Although acceptance levels are generally fairly high, studies have also identified various concerns with treatment use. Concerns with treatments include the potential for an escaped prescribed burn (generally the greatest concern for use of prescribed fire), increased prevalence of smoke, increased erosion, reduced water quality, impacts to wildlife or esthetics, and concern that mechanized thinning treatments may be used to promote commercial harvesting. It should be noted that concerns were not universal across studies, and treatments were considered as often for their potential positive impacts as negative impacts. For instance, in a survey of Northern Michigan residents, Kwon et al. (2008) found that participants believed that prescribed fires would improve wildlife habitat. Similarly, Vining and Merrick (2008) found that some respondents thought prescribed fires posed safety risks while others thought that they would reduce the safety risk.

Findings from a Utah study (Brunson and Evans, 2005) that resurveyed respondents who had been directly impacted by an escaped prescribed burn illustrate the complexity of treatment acceptance. While a high percentage of participants indicated that the escaped burn had negatively influenced their views about the use of prescribed fire, actual acceptance ratings remained constant across the study period (with ~80 percent indicating acceptance of some amount of prescribed fire use). However, other important changes emerged—participants expressed less confidence in forest managers to use prescribed fire effectively, were more concerned about fire use within 10 miles of their home, and also indicated more concern about the potential impacts of smoke on public health. Despite these increased concerns with smoke impacts, few participants (13 percent) indicated that prescribed fire should no longer be used due to the increased prevalence of smoke.

Findings from a number of studies suggest that, for a majority of the population, smoke is not a significant barrier to the use of prescribed fire and that a desire to improve forest health and/or reduce future fire risk tends to outweigh smoke concerns. However, findings also suggest that for a sizeable portion of the population—roughly a third of households—smoke is a major issue due to health concerns. For this segment of the population, smoke is likely a more dominant concern because of its implications for the health and well-being of family members. For individuals who are potentially affected, understanding how smoke issues are addressed in fire and fuel management efforts will be a highly salient issue.

2.3.1.2 Factors Influencing Treatment Approval

The two variables most frequently associated with fuel treatment acceptance are knowledge of a practice and trust in managers to implement it. The most common predictor of treatment acceptance across studies is the knowledge of, or familiarity with the practice. Some studies have also examined the influence of public outreach and education programs on treatment acceptance. Findings suggest that outreach programs can have a positive influence on knowledge and, in some cases, on attitudes toward treatments. Not all outreach programs are equally effective; results indicate that the success of outreach activities is influenced by both the quality of the content provided and the method by which it is communicated. Overall, interactive formats tend to be more highly rated.

Some studies have also found that higher knowledge levels about a treatment are associated with decreased concerns, particularly for prescribed fire. In Massachusetts, participants who self-reported having "some" or "a great deal" of knowledge were less concerned about effects of prescribed fire on esthetics and impacts to wildlife and their habitat (Blanchard and Ryan, 2007). The same study found that respondents on Long Island, who were more familiar with prescribed fire, were more willing to allow its use on private lands (Ryan and Wamsley, 2008). In Nevada, McCaffrey (2004) found that those who had read prescribed burning educational materials were more likely to think the practice improved wildlife habitat and diversity and less likely to agree that they did not like the appearance afterward or that smoke caused problems for a member of their household.

Studies have also found that citizen trust in management agencies significantly influences treatment acceptance. Across this research, trust has been conceptualized in different ways; common definitions describe trust as perceived competency of agency managers to implement treatments, perceptions of shared values between public participants and agency managers, or a combination of these two approaches. For example, Toman et al. (2011) found confidence in agency managers to effectively implement specific treatments (perceived competency) had the strongest influence on treatment acceptance, even when accounting for other variables (e.g., residency status, ratings of agency management, and general trust in agency managers).

2.3.2 Mitigation on Private Land

The second main focus of social science research to date is in relation to dynamics around homeowner mitigation on private property. This work has found that residents in fire-prone communities are generally aware of their fire risk, and most report taking some action to reduce that risk. These findings are consistent across studies and locations in the South, Northeast, Lake States, Rocky Mountains, Southern California, and the Pacific Northwest of the United States. Private landowners have implemented a range of practices, often recommended through FireWise and other programs, to mitigate their fire risk, including modifying vegetation, reducing flammability of structures, and developing an evacuation plan. However, not all activities are uniformly adopted; not surprisingly, activities with lower initial cost (either financial or in required time/effort) are more likely to be adopted. Although few studies have examined whether these actions are being maintained over time, what data there is suggests that property owners see their risk-reduction behaviors as a multiyear process, often discussing ideas about additional activities to complete in the future, and that many activities—such as raking needles, mowing vegetation adjacent to their homes, and clearing needles and leaves from their roofs—are seen as part of normal outdoor chores.

2.3.2.1 Factors Influencing Adoption of Mitigation Measures

Wildfire studies, as well as research on other natural hazards, demonstrate that while having an awareness of fire risk is important, it does not automatically lead to adoption of risk-reduction behaviors. Studies have found that adoption of wildfire mitigation measures is influenced by both personal/psychological factors and situational characteristics such as conditions of adjacent properties and residency status. While both types of factors have been found to influence decisions, there is evidence that the former are more influential. Personal/psychological factors that influence adoption of mitigation measures include perceived effectiveness of risk-reduction activities, self-efficacy (belief in their ability to complete treatments), and, for some WUI residents, perceived norms (e.g., beliefs about the attitudes of others toward treatments).

Several situational characteristics may influence adoption of risk reduction measures. Local ecological conditions are a consideration for many residents who have indicated a greater likelihood of adopting treatments they view as appropriate to the local ecological context. Residents also recognize that their risk is influenced by conditions on adjacent lands. In some cases, studies found that this provides motivation to reduce fuels on their properties to be a good neighbor and to do their part to contribute to shared protection. However, in other locations, residents have indicated they are unlikely to adopt risk-reduction behaviors on their properties because they believe that they would be ineffective given the poor condition of neighboring properties, including adjacent public lands. This recognition of shared risks has prompted some communities to adopt cooperative, communitywide risk-reduction efforts. While such efforts were effective at influencing behavior in those locations, community-organized programs were not needed elsewhere as homeowners worked individually or directly with adjacent neighbors to take action on their properties and across property boundaries.

Residency status (whether residents were part-time or full-time residents) is a final situational factor that may influence treatment adoption. While some studies found few differences between seasonal and permanent residents, others found that full-time residents had more positive attitudes toward, or were more likely to adopt risk-reduction behaviors, particularly the more involved treatments such as tree removal. Findings suggest that the time required to undertake mitigation measures can be was particularly important to part-time residents, who have indicated that they did not want to spend their limited time at their properties engaged in such activities. Conforming with neighborhood norms may be a more important factor for permanent residents. Absentee landowners who never or rarely visited their properties were more likely to be disconnected from the local situation and take few fire preparedness actions.

Overall, studies have shown that residents attempt to balance risk-reduction behaviors with other values they hold for their properties, such as privacy, perceived naturalness, shading, wildlife habitat, and potential esthetic impacts (although esthetic improvements were also often cited as rationale for adopting risk reduction measures WUI residents weigh the expected risk-reduction benefits of a mitigation measure with the potential impacts on these other values and, in some cases, make decisions such as leaving shrubs to provide screening from neighboring properties or leaving trees to provide views from windows even though they understand this may increase their fire risk. In addition to the perceived tradeoffs between risk-reduction behaviors and other values people hold for their properties, residents across locations most frequently cited financial cost and time constraints as barriers to implementation. In some locations, residents also noted the challenging nature of the work and indicated an inability to complete the work as a significant barrier. This perception was driven by physical limitations, a lack of knowledge about what specifically should be done at the property level, or a lack of necessary equipment.

2.3.2.2 Mitigation Responsibility

When asked about who is responsible for undertaking risk reduction measures, most residents view mitigating fire risk on their property as their responsibility. However, in recognition that their risk is influenced by the condition of adjacent lands, residents see the responsibility for mitigation as shared; each landowner, whether private or public, is responsible for mitigation of the fire risk for their property. Although residents did not see the government as having a mitigation responsibility on private land, they did support the idea that government agencies had some responsibility for providing educational materials and, in some cases, technical assistance to help homeowners understand local fire conditions and specific methods to mitigate their fire risk. While multiple methods can be used to provide such information, several studies indicate that interactive methods are particularly effective.

2.3.3 Working with Communities

Only a few studies have examined the dynamics of collaboration in the context of wildfire. In 15 case studies of wildfire planning and preparedness conducted throughout the country, Sturtevant and Jakes (2008) found that collaboration was integral to successful wildfire risk planning at the community level. Another study examined the Fire Learning Network, which is designed to link local level collaborative groups into larger regional and national networks interested in restoring fire adapted landscapes (Goldstein and Butler, 2010). Leaders of the local collaborative groups meet periodically with regional partners to share successes and mistakes, receive peer reviews of their restoration plans, and build expertise. A review has found that the network has successfully contributed to the development of local expertise while supporting local collaborative efforts (Goldstein and Butler, 2010). Early studies on the effectiveness of developing Community Wildfire Protection Plans (CWPP) in reducing wildfire risk find mixed results. One project found a lack of innovation in fire management approaches in examined CWPPs (Brummell et al., 2010). Another study reviewed the development and implementation of two CWPPs in Oregon. At the time of the study, the Forest Service had elected not to implement the fuel reduction plans in either one, but for different reasons. In one case, the Forest Service had not participated in the development of the CWPP and did not choose the CWPP prescription in their planning process, and in the other case, insufficient funds in the USFS budget was attributed to nonimplementation (Fleeger and Becker, 2010). However, a different study that surveyed state-level wildfire program managers in 11 states found that CWPPs were rated as one of the more effective elements for the overall success of programs designed to mitigate risk on private land (Renner et al., 2010).

2.4 DURING AND POSTFIRE SOCIAL DYNAMICS

Although most social science research in the United States to date has focused on prefire concerns, there has been growing research interest in social dynamics during and after fires. Although more limited, this body of work provides a sense of key variables to consider for future management efforts, as well as for future research.

2.4.1 During-Fire Considerations

2.4.1.1 Evacuation

Fires that directly threaten a community can lead to substantial psychological, physical, and financial impacts. While designed to limit loss of life, evacuations themselves can result in significant stress and social disruption to residents. Evacuated residents indicate substantial anxiety over the status of their home and property and a lack of control of ongoing events. Limited research suggests that homeowner decisions to evacuate are influenced by the nature of the evacuation order (e.g., mandatory vs voluntary), the fire readiness of their home and property, previous evacuation experiences, and complicating factors such as ownership of pets and livestock, age, and health of family members, etc.

Although evacuation is seen as the surest way to ensure human safety and is the most common response in the United States, the evacuation process is not without risks. For this and other reasons, managers and researchers have begun to consider alternatives to the evacuation of residents during wildland fire events. The "Shelter in Place" (SIP) model has been used during other disasters in the United States, whereas the "Stay and Defend or Leave Early" approach is commonly used in Australia. The limited available research suggests that successful adoption of either alternative in the United States will require a substantial shift in the paradigm of fire management for both residents and fire management personnel. For example, in one of the few US locations where alternative plans have been developed (communities developed to SIP in Southern California), the local fire community disagreed on the definition of SIP and whether it should be used as a primary response or a last-ditch effort only if evacuation was not possible. At the same time, most residents in the designated communities did not know what to do should a fire occur (Paveglio et al., 2010a). However, research in a rural Idaho community suggested that alternatives to evacuation could be viable in certain circumstances and with appropriate preparation (Paveglio et al., 2010b), indicating the issue warrants further exploration.

2.4.1.2 Communication

During a fire, residents seek real-time information to help them decide on appropriate behaviors. In the initial stages, residents seek information about the fire location, when and how an evacuation order will be issued, and details about available services (e.g., location of shelters, availability of support to transport and board pets/livestock, where additional information can be obtained). Once evacuated, residents want to know how the fire has affected their homes and places they care about. When the fire no longer is seen as posing a significant threat to the community, resident information needs shift to learning when they will be allowed to return home, remaining health and safety risks, and the availability of services to help them in their recovery efforts (e.g., grief counseling, insurance, disposal of burned material, and rebuilding assistance).

Throughout a fire, residents are likely to draw on multiple information sources to address their information needs. Mass media sources are generally seen cited as being overly sensational and providing inaccurate information. A recent study found an expanding use of informal sources and social media including local web sites, blogs, internet-based forums, and mobile phones (Sutton et al., 2008). In some locations, studies have identified a tension between the information needs of residents, who may seek near continuous, specific information during a fire, and agency practices that may delay information to ensure quality control or emphasize delivery of tactical information (e.g., size of fire and resources dedicated to fire protection).

2.4.2 Postfire Considerations

2.4.2.1 Impacts

Wildland fires are a social as well as an ecological disturbance and have the potential to have far reaching impacts on the surrounding communities. Some impacts are tangible, such as damaged homes and infrastructure, and potential flooding issues, while many other impacts may be less obvious, but no less significant, ranging from the stress of evacuation to grief over changes to the surrounding landscape. Similar to what has been found with other natural hazards, experiencing a fire can lead to a variety of long-term responses. For some individuals, the experience will increase motivation to take proactive risk-reduction measures, while others may be less likely to engage in risk-reduction behaviors due to a sense of fatalism (e.g., seeing risk reduction efforts as ineffective after witnessing loss of homes that had implemented mitigation activities) or a belief that such behaviors are unnecessary because they believe local conditions have changed enough that fire is less of a threat.

In addition to individual level impacts, wildfires can also result in changes at the community level. In some locations, residents have reported an increased sense of community as residents, local businesses, and agency personnel worked together during and immediately after the fire event to help each other and to protect their homes and valued natural resources. However, in other cases, disputes about how the fire was managed, particularly underutilization of local firefighting resources, or over appropriate land management prior to the fire, may negatively affect agency—community relationships after the fire.

The postfire landscape presents new management challenges. Research indicates that there are high levels of support for many postfire management activities. Immediate postfire stabilization activities, such as erosion control, and removal of hazard trees, particularly along trails and in other public areas, have been found to have high levels of support. Broader forest management decisions, such as salvage logging and restoration actions, tend to have a greater range of opinions. Support for either can be high under appropriate conditions. The level of support can depend on location, values placed on the trees (economic or ecological), and the perceived risk to the forest with intervention or nonintervention. Most studies have reported finding preference for a balanced approach; take some burned trees in order to not waste them and recoup some economic value, and also leave some standing dead trees for wildlife and shade for seedlings. Support for harvesting has also been found to be correlated with levels of trust that citizens have in the implementing agency, with how the fire was managed, and handling of postfire decision making.

2.4.2.2 Communication and Outreach

Research has begun to identify a number of factors that contribute to successful postfire outreach. Research has shown several areas of interest to the public after a fire event including cause of the fire, how it could have been prevented, goals and reasons for postfire management actions, and outcome of restoration efforts. How messages are communicated is also important, particularly the need for two-way communication, including having agency personnel ask for and utilize forest-users knowledge and experience in the local area. Field tours have been shown to be an effective means of increasing the understanding of forest and fire ecology, what happened during the fire, and options for postfire management. Where field tours are not possible, visual presentations at public meetings with photographs of burned sites have aided in the understanding of the complexities of postfire management.

Fire events may inspire local citizens to participate in fire recovery efforts, through planning or on-the-ground restoration activities. When agencies are willing to engage citizens and offer opportunities for them to participate in restoration efforts, citizens have reported improved relationships with agency personnel. Perhaps more importantly, many citizens have reported that participating in on-the-ground restoration activities after a fire helped them to reconnect with the forest and to heal from the fire. These efforts have been most successful when projects are located in locally important areas such as popular recreation spots or view sheds.

2.5 GEOGRAPHIC AND SOCIODEMOGRAPHIC DIFFERENCES

It is commonly believed that individuals living in different regions of the country or with different sociodemographic characteristics will respond differently to fire management issues. However, analysis of social science research findings from 2000–2010 indicates that geographic and socio-demographic differences are rarely key explanatory factors where fire management knowledge, attitudes, or actions are concerned.

Studies have been conducted at sites throughout the United States, and many have explicitly included geographic variation as part of their design. Notably, the most consistent finding across these studies is that they detected much less variation than expected. Where geographic variation has been found, it either has generally been too small to be meaningful or was seen to reflect specific local contextual factors, such as ecological conditions, regulations, building styles, agency–community interaction, or specific historical events. This is not to discount such differences when they exist; such differences can be highly influential highlighting why managers need to understand their local communities and tailor programs to the context.

When discussing sociodemographic factors, studies address two general categories—standard demographic measures (age, income, education level, and gender) and residential characteristics, such as length of residence and type of residency (permanent or seasonal). The most apparent dynamic for both these measures is how often these variables are found to have no significant relationship with key variables, particularly support or approval of a treatment. Of the few studies that report significant relationships between a socio-demographic variable fire-related attitudes and behavior, relationships are not consistent among studies, and no meaningful pattern can be identified for all but two sociodemographic variables: type of residency and gender. As noted earlier, some studies have found that part-time residents are less likely to undertake mitigation activities, particularly more time-consuming actions. Gender differences have most commonly been found in relation to risk response finding that women have a higher risk perception and concern levels and lower support for more controversial practices such as prescribed fire and herbicides. For both residency and gender, however, it is important to note that the majority of studies either do not report on the variables or find no significant relationship with fire related attitudes and behaviors. One reason why sociodemographic variables may have such a limited influence is suggested by Absher and Vaske (2006) who found that psychological variables (familiarity, effectiveness, and esthetics) explained substantially more variance in approval of fuel treatments and the likelihood of taking mitigation measures on one's property than demographic variables.

2.6 CONCLUDING REMARKS

The current wildfire challenge in the United States is, in part, a product of ecological changes created by decades of fire suppression and climate change. It is greatly complicated by the increasing social complexity created by the growing number and diversity of organizations and people impacted by wildfires. This has created a situation where the traditional fire management approach no longer effectively mitigates the fire risk, and new approaches are needed. Given the increased social complexity, a key component of identifying new approaches will be in understanding how diverse social dynamics are affected by, and in turn can influence, wildfire outcomes. Ultimately reducing fire risk is not about eliminating fire; fire will occur on the landscape. However, communities and individuals can reduce the risk of negative consequences when wildfires do occur. In the face of a changing environment and as more people move into natural areas, the current body of knowledge described here, along with future research findings, will be increasingly relevant to development of fire adapted communities.

This chapter has provided an overview of key findings from the past decade of fire social science research. Results indicate that many oft-heard descriptions of the public as not understanding the fire risk, not taking responsibility for mitigating that risk on their land, or not supporting fuels treatments on public lands are not accurate. Instead, on the whole, WUI residents understand their fire environment, support fuels treatments on public lands, and are undertaking mitigation actions on their property. Most residents in fire-prone areas not only understand the fire risk but also the ecological role of fire. The vast majority of the public support some amount of prescribed fire and mechanized thinning with knowledge of a practice and trust in those implementing it key in shaping acceptance. These findings, combined with findings that "no action" is consistently the least preferred alternative, suggest that there is greater public support for active rather than passive management in achieving fire risk reduction goals. Further, WUI residents believe that it is their responsibility to reduce fire risk on their property. Many are taking action, but the decision process to act is complex; property owners balance their fire risk with other values they hold for their properties, considerations of potential efficacy of the action, and their ability to implement it. Property owners are more likely to adopt those behaviors they perceive as compatible with their other values as well as those they believe will provide enough benefits to outweigh costs.

While there is less research on social dynamics during and after fires, the work points to several important considerations, for managers and for future research. Experiencing a fire is a stressful process, particularly evacuation.

Timely and accurate communication during this period is particularly important to minimizing the uncertainty caused by a fire. However, there is a need for better understanding how best to improve outcomes, whether in terms of communication and the evacuation process or in terms of when evacuation is or is not the most appropriate response. Experiencing a wildfire can have diverse long-term impacts on a community whether in terms of loss of homes, loss of a valued landscape, or agency—community relationships. Findings suggest there are high levels of support for many postfire management activities, including salvage logging under appropriate conditions. Open communication and, when possible, including citizens in on-the-ground postfire recovery effort provide ways to see and understand the effects of the fire, share perspectives with agency personnel, and can provide a tangible way to participate in the forest's recovery, which in turn can help with their own recovery.

A thread running through the findings is that effective communication and outreach are important throughout the fire management process—before, during, and after an event—and that interactivity is a key component of effective communication. Outreach programs and citizen—agency interactions before an event can help residents both to understand management efforts on public lands and to also help them identify and implement mitigation measures on private property. During and after a fire, residents have an ongoing need for information on the fire and its impacts on their home and property and on places they care about; this information helps reduce the stress and anxiety associated with the uncertainty of experiencing such a disruption to normal lives. Communities that reported being well informed by fire agencies during and after a wildfire event have tended to experience less negative emotion during the fire and less postfire stress.

A second thread running through the findings is that actions taken at one stage can have a lasting effect, positively or negatively, on public response and citizen—agency interactions at another stage. Understanding and trust that are built at one point can facilitate ease of exchange and support at another stage. Perceptions of how forest management decisions were made and implemented before a fire can influence views both of how a fire was managed and on acceptable postfire management actions. Trust in agency personnel has been found to be correlated with acceptance of prefire treatments such as prescribed fire and with postfire treatments such as salvage logging. Agency efforts to connect with local groups during a fire event can lead to the development of partnerships with local governments and local citizens to address postfire recovery, landscape restoration, and to prepare for future fire events. Perceptions that a fire was well managed can lead to increased community cohesion and strengthened agency ties following a fire event. However, the opposite can also be true; when residents perceive that a fire or immediate postfire phase is poorly managed, this can lead to reduced confidence in agency managers or acceptance of management activities. Where communities and agencies have sufficiently prepared, recovery from a wildfire event is likely to proceed more smoothly than in places where little or no prefire planning has taken place.

Collectively, this body of research demonstrates that individuals, communities, policy makers, and fire management agencies are working to create fire adapted communities. However, there is still much work to be done, in both the research arena and on-the-ground activities. Ultimately, citizens' attitudes, confidence in agency managers, and acceptance of agency activities are linked across the different phases of a fire event. Recognizing these linkages can help managers take into consideration how actions taken at one point in time may affect outcomes and relationships down the road. Although before, during, and after a fire provides a convenient structure for discussing fire management, it will be important to more explicitly recognize and work across these stages.

REFERENCES

Absher, J.D., Vaske, J.J., 2006. An analysis of homeowner and agency wildland fire mitigation strategies. In: Peden, J.G., Schuster, R.M. (Eds.), Proceedings of the 2005 Northeastern Recreation Research Symposium, April 10-12, 2005, Bolton Landing, NY; GTR-NE-341. USDA Forest Service, Northeastern Research Station, Newtown Square, PA, pp. 231—236.

Behan, R.W., May 1975. Forestry and the end of innocence. Am. For. 16—19, 38—49.

Blanchard, B., Ryan, R.L., 2007. Managing the wildland—urban interface in the northeast: perceptions of fire risk and hazard reduction strategies. North. J. Appl. For. 24 (3), 203—208.

Bowker, J.M., Lim, S.H., Cordell, H.K., Green, G.T., Rideout-Hanzak, S., Johnson, C.Y., 2008. Wildland fire, risk, and recovery: Results of a national survey with regional and racial perspectives. J. of Forestry 106 (5), 268—276.

Brummel, R.F., Nelson, K.C., Souter, S.G., Jakes, P.J., Williams, D.R., 2010. Social learning in a policy-mandated collaboration: community wildfire protection planning in the eastern United States. J. Environ. Plann. Manage. 53 (6), 681—699.

Brunson, M.W., Evans, J., 2005. Badly burned? Effects of an escaped prescribed burn on social acceptability of wildland fuels treatments. J. For. 103 (3), 134—138.

Brunson, M.W., Shindler, B.A., 2004. Geographic variation in social acceptability of wildland fuels management in the western United States. Soc. Nat. Res. 17 (8), 661—678.

Davis, C., 2001. The West in flames: the intergovernmental politics of wildfire suppression and prevention. Publius 31 (3), 97—110.

Davis, C., 2006. Western wildfires: a policy change perspective. Rev. Policy Res. 23 (1), 115—127.

Fleeger, W.E., Becker, M.L., 2010. Decision processes for multi-jurisdictional planning and management: community wildfire protection planning in Oregon. Soc. Nat. Res. 23 (4), 351—365.

Goldstein, B.E., Butler, W.H., 2010. Expanding the scope and impact of collaborative planning. J. Am. Plann. Assoc. 76 (2), 238—249.

Gorte, R.W., 2003. Policy response. In: Cortner, H.J., Field, D.R.J., Pamela, J., Buthman, J.D. (Eds.), Humans, Fires, and Forests: Social Science Applied to Fire Management. Workshop summary Ecological Restoration Institute, Flagstaff, AZ, pp. 59—63.

Huntsinger, L., McCaffrey, S., 1995. A forest for the trees: forest management and the Yurok environment, 1850−1994. Am. Indian Cult. Res. J. 19 (4), 155−192.

Hays, S.P., 1959. Conservation and the Gospel of Efficiency: The Progressive Conservation Movement. Atheneum, New York, 1890−1920.

Jensen, S.E., 2006. Policy tools for wildland fire management: principles, incentives, and conflicts. Nat. Res. J. 46 (4), 959−1003.

Kneeshaw, K., Vaske, J.J., Bright, A.D., Absher, J.D., 2004. Acceptability norms toward fire management in three national forests. Environ. Behav. 36 (4), 592−612.

Kwon, J., Vogt, C., Winter, G., McCaffrey, S., 2008. Forest fuels treatments for wildlife management: do local recreation users agree? In: LeBlanc, C., Vogt, C. (Eds.), Proceedings of the 2007 Northeastern Recreation Research Symposium, April 15−17, 2007, Bolton Landing, NY; GTR-NRS-P-23. USDA Forest Service, Northern Research Station, Newtown Square, PA, pp. 132−137.

Lewis, D.R., 1993. Still native: the significance of native Americans in the history of the twentieth-century American west. The West. Hist. Q. 24, 203−227.

McCaffrey, S., 2004. Fighting fire with education: what is the best way to reach out to home-owners? J. For. 102 (5), 12−19.

McCaffrey, S., Moghaddas, J.J., Stephens, S.L., 2008. Different interest group views of fuels treatments: survey results from fire and fire surrogate treatments in a Sierran mixed conifer forest, California, USA. Int. J. Wildland Fire 17 (2), 224−233.

McCaffrey, S., Olsen, C., 2012. Research Perspectives on the Public and Fire Management: A Synthesis of Current Social Science on 8 Essential Questions. Gen. Tech. Rep. NRS-104. Dept of Agriculture, Forest Service, Northern Research Station, Newtown Square, PA, U.S., 40 pp.

Moseley, C., 2007. Class, ethnicity, rural communities, and the socioeconomic impacts of fire policy. In: Daniel, T.C., Carroll, M.S., Moseley, C., Raish, C. (Eds.), People, Fire, and Forests: A Synthesis of Wildfire Social Science. Oregon State University Press, Corvallis, OR, pp. 171−186.

Paveglio, T.B., Carroll, M.S., Jakes, P.J., 2010a. Adoption and perceptions of shelter-in-place in California's Rancho Santa Fe Fire Protection District. Int. J. Wildland Fire 19 (6), 677−688.

Paveglio, T.B., Carroll, M.S., Jakes, P.J., 2010b. Alternatives to evacuation during wildland fire: exploring adaptive capacity in one Idaho community. Environ. Hazards 9 (4), 379−394.

Pyne, S., 1997. Fire in America: A Cultural History of Wildland and Rural Fire. University of Washington Press, Seattle.

Renner, C.R., Haines, T.K., Reams, M.A., 2010. Building better blocks. Wildfire 19 (2), 10−16.

Ryan, R.L., Wamsley, M.B., 2008. Public perceptions of wildfire risk and forest management in the Central Pine Barrens of Long Island (USA). Australas. J. Disaster Trauma Stud. 2008 (2).

Shindler, B.A., Toman, E., 2003. Fuel reduction strategies in forest communities. J. For. 101 (6), 8−15.

Shindler, B.A., Toman, E., McCaffrey, S.M., 2009. Public perspectives of fire, fuels, and the Forest Service in the Great Lakes Region: a survey of citizens in Minnesota, Wisconsin, and Michigan. Int. J. Wildland Fire 18 (2), 157−164.

Shindler, B.A., Gordon, R., Brunson, M.W., Olsen, C., 2011. Public perspectives of sagebrush ecosystem management in the Great Basin. Rangeland Ecol. Manage. 64 (4), 335−343.

Steelman, T.A., Burke, C.A., 2007. Is wildfire policy in the United States sustainable? J. For. 105 (2), 67−72.

Steelman, T.A., Kunkel, G., Bell, D., 2004. Federal and state influence on community responses to wildfire threats: Arizona, Colorado, and New Mexico. J. For. 102 (6), 21−27.

Stephens, S.L., Ruth, L.W., 2005. Federal forest-fire policy in the United States. Ecol. Appl. 15 (2), 532−542.

Sturtevant, V., Jakes, P., 2008. Collaborative planning to reduce risk. In: Martin, W.E., Raish, C., Kent, B. (Eds.), Wildfire Risk: Human Perceptions and Management Implications. Resources for the Future, Washington, DC. pp. 44−63.

Sutton, J., Palen, L., Shklovski, I., 2008. Backchannels on the front lines: Emergent uses of social media in the 2007 Southern California wildfires. In: Fiedrich, F., Van de Walk, B. (Eds.), 5th International Conference on Information Systems for Crisis Response and Management ISCRAM2008, May 5−7, 2008. Washington, DC, USA, pp. 624−631.

Toman, E., Stidham, M., Shindler, B., McCaffrey, S., 2011. Reducing fuels in the wildland urban interface: community perceptions of agency fuels treatments. Int. J. Wildland Fire 20 (3), 340−349.

Toman, E., Stidham, M., McCaffrey, S., Shindler, B., 2013. Social Science at the Wildland Urban Interface: A Compendium of Research Results to Create Fire Adapted Communities. Gen. Tech. Rep. NRS-111. U.S. Department of Agriculture, Forest Service, Northern Research Station, Newtown Square, PA, 75 pp.

Vining, J., Merrick, M.S., 2008. The influence of proximity to a National Forest on emotions and fire-management decisions. Environ. Manage. 41 (2), 155−167.

Winter, P.L., 2002. Californians' opinions on wildland and wilderness fire management. In: Jakes, P.J. (Ed.), Homeowners, Communities, and Wildfire: Science Findings from the National Fire Plan. Proceedings from the Ninth International Symposium on Society and Resource Management, June 2−5, 2002, Bloomington, Indiana. GTR-NC-231, pp. 84−92. St. Paul, MN.

Westerling, A.L., Hidalgo, H.G., Cayan, D.R., Swetnam, T.W., 2006. Warming and earlier Spring increase western U.S. Forest wildfire activity. Science 313, 940−943.

Chapter 3

Wildfire: A Canadian Perspective

Tara McGee
Department of Earth and Atmospheric Sciences, University of Alberta, Edmonton, AB, Canada

Bonita McFarlane
Natural Resources Canada, Canadian Forest Service, Northern Forestry Centre, Edmonton, AB, Canada

Cordy Tymstra
Alberta Environment and Sustainable Resource Development, Forestry and Emergency Response Division, Wildfire Management Branch, Edmonton, AB, Canada

ABSTRACT
This chapter provides a broad overview of the ecological and social aspects of wildfires, their impacts, and management in Canada. We describe the Canadian fire landscape, including vegetation and fire ecology, with an emphasis on the boreal forest. Wildfire causes and their impacts from past to recent events in 2003 and 2011, and communities at current and future risk of wildfires are described. Wildfire management, including land use, institutions involved in wildfire management, and policies and practices are explained. Wildfire mitigation as practiced by governments and homeowners, community response to wildfire, and relief and recovery following a wildfire are discussed.

3.1 INTRODUCTION

Canada has 10 percent of the world's forests and 30 percent of the world's boreal forest within its borders (Natural Resources Canada, 2012). The Canadian boreal forest is one of the largest ecosystems on earth and is the largest forest region in Canada, accounting for 54 percent of the country's 347.7 million hectares of forestland (National Forest Inventory, 2006). Forests are an important resource for the Canadian economy and society (Natural Resources Canada, 2012; Rowe, 1992). In 2011, the forest product sector contributed $23.7 billion to the economy (1.9 percent of the gross domestic product) and was the main economic driver for about 200 rural communities

Wildfire Hazards, Risks, and Disasters. http://dx.doi.org/10.1016/B978-0-12-410434-1.00003-8

(Natural Resources Canada, 2012). Canada's forests provide many additional benefits, including ecosystem services, such as clean air and water, biodiversity, and carbon storage, nontimber forest products, and recreation (Anielski and Wilson, 2009), and are home to 80 percent of Canada's Aboriginal communities (Natural Resources Canada, 2012). The forested landscape is integral to Canadian culture, inspiring world-renowned painters such as the Group of Seven and Emily Carr.

Wildfire is a natural part of the Canadian forest landscape. Indeed, Pyne (2007a, p. 960) describes Canada as "a large and combustible swathe of fire-planet Earth. Historically, fires swept its prairies every two or three years; combusted its Cordilleran forests every five to fifty; and devoured its boreal forest, in immense chunks, every 50—120 years…." On average, about 8,400 wildfires consume over two million ha of forested lands each year. Large fluctuations in area burned occur with some years exceeding four million hectares (Figure 3.1).

This chapter explores the Canadian wildfire situation and the social and institutional dynamics that underlie wildfire risk management. Specifically, we describe the Canadian fire landscape including a brief overview of vegetation and fire ecology. This sets the stage for a description of wildfires in Canada, land use, and institutions involved in wildfire management. Finally, we discuss wildfire management policies and practices, wildfire mitigation by governments and homeowners, community response to wildfire, and relief and recovery from wildfire.

3.1.1 Wildfire and Vegetation

Canada's diverse landscapes and ecosystems support a rich biodiversity of >700,000 species of flora and fauna (Commissioner of the Environment and Sustainable Development, 2013). Several classification schemes have been developed to describe Canada's ecological diversity. The national ecological framework for Canada incorporates 15 terrestrial ecozones; broad regions with similar landform, climate, wildlife, soil, and vegetation (Ecological Stratification Working Group, 1995). This classification, while useful for the spatial analysis of wildfire activity (Krezek-Hanes et al., 2011) does not directly account for wildfire as the predominant process shaping much of Canada's landscapes. To project regional risk and wildfire—climate interactions across Canada, Boulanger et al. (2012) proposed an alternate zonation based on homogeneous wildfire regimes. Rowe (1972) classified the forest cover into eight regions (Acadian Forest, Carolinian Forest, Coastal Forest, Great Lakes—St. Lawrence Forest, Columbia Forest, Montane Forest, Subalpine Forest, and the Boreal Forest) based on predominant tree species and stand types. We aggregate Rowe's eight regions into four (Boreal Forest Region; Montane, Subalpine, and Columbia Forest Region; Deciduous and Coastal Forest Region; and Great Lakes—St. Lawrence Forest Region) based on wildfire regime characteristics, and we use these as the basis for discussion in this chapter.

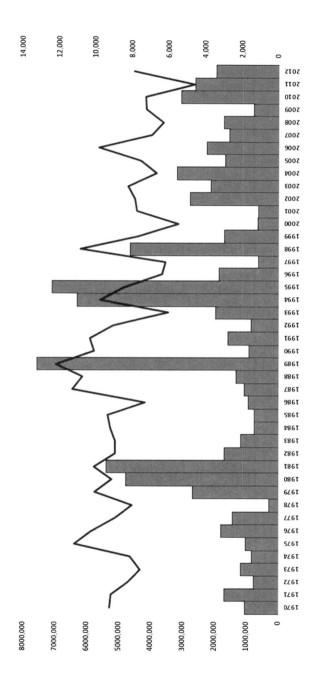

The Boreal Forest Region is the largest forest region in Canada, arcing in a broad band from Yukon in the northwest to Newfoundland in the east (Figure 3.2). Black spruce (*Picea mariana*), white spruce (*Picea glauca*), balsam fir (*Abies balsamea*), larch (*Larix laricina*), lodgepole pine (*Pinus contorta*), jack pine (*Pinus banksiana*), trembling aspen (*Populus tremuloides*), balsam poplar (*Populus balsamifera*), and white birch (*Betula papyrifera*) are the predominant trees species in the boreal forest. Although the number of tree species is relatively low, the diversity of understorey plant species in comparison is considerably greater (Roberts, 2004).

Most wildfire in Canada occurs in the Boreal Forest Region. Krezek-Hanes et al. (2011) reported that the area burned by large (>200 ha) wildfires in the boreal forest accounted for 92 percent of the total area burned in Canada between 1959 and 2007. A small percentage of wildfires (3 percent) are responsible for 97 percent of the burned area in the boreal forest (Stocks et al., 2002). Wildfires in Canada's boreal are characterized by infrequent, large (>200 ha), high-intensity, crown fires (de Groot et al., 2013). Liu et al., 2010 suggest an increasing fire season length and occurrence of extreme weather

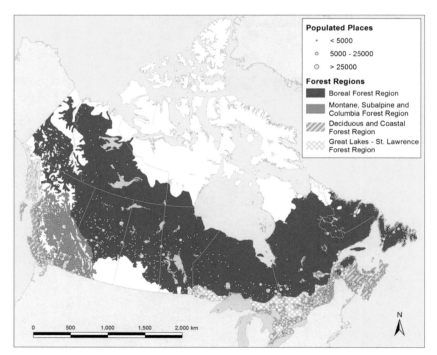

FIGURE 3.2   Canada's forest regions and populated places. © 2015 Bonita McFarlane. Published by Elsevier Inc. All rights reserved. *(Population data derived from Statistics Canada Census (2011) and forest regions derived from Rowe (1972).)* Communities outside of forested regions and communities that contain <20 percent forest within a 2.4-km buffer are excluded.

events are the most significant global impacts of climate change for many fire-dependent forest ecosystems. In Canada, predicted climate changes are expected to increase fire weather severity and area burned (Flannigan et al., 2009, 2013; Tymstra et al., 2007).

The expansiveness and biodiversity of the boreal forest contribute to its ecological significance. Pure deciduous, pure coniferous, and mixed deciduous–coniferous stands are intermingled with rivers, lakes, fens, bogs, and marshes. Peatland ecosystems are wetlands with peat accumulation to depths of ≥ 40 cm (Wells and Zoltai, 1985). Approximately 60 percent of the total boreal forest carbon sink is estimated to be stored in peatlands (Turunen, 2002). Future fire regimes under climate change may result in more and deeper burning in peatlands, resulting in a significant increase in carbon emissions (Turetsky et al., 2011a,b; Price et al., 2013).

Canada's other forest regions account for approximately 46 percent of the total forest cover. The predominant tree species in the Deciduous and Coastal Region include red spruce (*Picea rubens*), balsam fir, and yellow birch (*Betula alleghaniensis*) in eastern Canada; American beech (*Facus grandifolia*), sugar maple (*Acer saccharum*), black walnut (*Juglans nigra*), red hickory (*Carya glabra*), and red oak (*Quercus rubra*) in southwestern Ontario; and western red cedar (*Thuja plicata*), western hemlock (*Tsuga heterophylla*), Sitka spruce (*Picea sitchensis*), and Douglas-fir (*Pseudotsuga menziesii*) along the west coast in British Columbia. Wildfires in these forests are infrequent and relatively small and of a low-to-moderate intensity. Similar wildfires occur but more frequently in the Great Lakes–St. Lawrence Forest Region, which is characterized by red pine (*Pinus resinosa*), eastern white pine (*Pinus strobus*), eastern hemlock (*Tsuga canadensis*), yellow birch, sugar maple, and red oak. Historically, white and red pine stands experienced low-to-moderate intensity wildfires every 20–30 years. Cyclic wildfire is an important ecological process for the maintenance of pine ecosystems. Although infrequent, high-intensity wildfires can occur when this forest region experiences severe drought.

In British Columbia, the Montane, Subalpine, and Columbia Forest Regions account for approximately 80 percent of the province's 55 million hectares of forests. These forests are ecologically diverse dry douglas-fir, lodgepole pine, ponderosa pine (*Pinus ponderosa*), and aspen forests in the south interior (montane forests), western red cedar, western hemlock, and Douglas-fir forests in the southeast (Columbia forests). The subalpine forests (Engelmann spruce—*Picea engelmannii*, subalpine fir—*Abies lasiocarpa*, and lodgepole pine) occur in both British Columbia and Alberta. The forests in the Montane, Subalpine, and Columbia Forest Region experiences a mixed fire regime of infrequent large, high-intensity wildfires, and more frequent, low-to-moderate intensity wildfires, particularly in the montane forests.

The occurrence of destructive wildfires beyond Canada's forested regions is an increasing concern. Although the Canadian tundra is not exempt from wildfire ignitions from lightning and human activity, wildfires in the past have been relatively rare events (Wein, 1976). A reported increase in the frequency of large tundra fires, however, suggests these wildfires may increase the release of soil-bound carbon (Mack et al., 2011).

Wildfires in Canada's grasslands are also a concern. Despite most of Canada's native prairie grasslands being converted to croplands and pastures, diverse and ecologically significant grassland communities remain in southeast Alberta, south Saskatchewan, and southwest Manitoba (Shorthouse, 2010). Dispersed fescue-wheatgrass grasslands occur in British Columbia, and in southwest Yukon, and scattered grasslands occur on dry, south-facing slopes.

Fescue (*Festuca campestris*) grasslands cover approximately 1.3 million hectares predominantly along the foothills in southwest Alberta (Shorthouse, 2010). As the moisture regime transitions from the west to the east and the north to the south, so do the grassland communities. To the east of the fescue grasslands lies a circular band of moist, mixed grasslands characterized by dense communities of various grasses (e.g., western porcupine grass—*Stipa curtiseta*, needle-and-thread grass—*Stipa comata*, western wheat grass— *Agropyron smithii*, and northern wheat grass—*Agropyron dasystachyum*) and shrubs (e.g., western snowberry—*Symphoricarpos occidentalis*, chokecherry— *Prunus virginiana*, Saskatoon—*Amelanchier alnifolia*, wolf willow— *Elaeagnus commutata*, and wild rose—*Rosa acicularis*). Semiarid mixed grasslands lie to the south of the moist mixed grasslands. Short to midsized grasses (needle-and-thread grass, blue grama—*Bouteloua gracilis*, western wheat grass, and June grass—*Koeleria macrantha*) and dwarf sedges dominate this landscape. Extending north from the United States are fragmented tall grass prairie grasslands in southwest Manitoba. The dominant species include big bluestem (*Andropogan gerardi*), Indian grass (*Sorghastrum nutans*), little bluestem (*Andropogan scoparius*), prairie dropseed (*Sporobolus heterolepsis*), northern wheat grass, and Canada wild rye (*Elymus Canadensis*). Without periodic wildfires, trees and shrubs invade the fescue, mixed, and tall grasslands.

Wildfires are integral to grassland ecosystem conservation (Wright and Bailey, 1982), but can impact farmland, communities, and infrastructure, and present challenges to rural fire services. In 2013, the 5,056-ha Grasslands National Park wildfire in Saskatchewan destroyed straw stacks, heritage buildings, trail markers, signage, and visitor infrastructure (PWSS, 2013). In January 2012, grassland wildfires destroyed about 30 barns, sheds and workshops, fences, corrals, and four homes, and killed livestock and injured firefighters in Alberta (The Pincher Creek Voice, 2012). In 2011, two large grassland fires occurred in the same area forcing the evacuation of an estimated 3,500 residents (The Globe and Mail, 2012). Advanced warnings of forecasted extreme, fire-danger conditions, early detection, and rapid, coordinated initial attack, are important for the effective control of all wildfires, but particularly so for grassland fires. This is because land owners and other residents in farming communities may decide to combat these fires despite having little or no training, or communication and personal protective equipment. When sudden changes in wind speed and direction change the rate and direction of spread of the grassland fires, these volunteers are confronted with very dangerous conditions. Increasing the public's awareness of grassland fire-behavior potential (they can spread faster than forest fires) may help to reduce wildfire starts and mitigate losses when they do occur (Alexander et al., 2013).

3.1.2 Boreal Forest Wildfire Ecology

Wildfire plays a crucial role in the ecology of northern boreal forests (Wright and Heinselman, 1973). The thin humus layers, ash layers, and exposed mineral soil created by wildfire provide favorable seedbeds. The consumption of organic matter increases available nutrients essential for plant growth. Competition for water and nutrients is also reduced. The high temperatures produced by wildfires stimulate plant sprouting and trigger the release of seeds from aerial and ground seed banks for some species.

Wildfire influences the structure and composition of forest stands (Wright and Heinselman, 1973). By controlling stand age classes and successional rates, wildfire shapes the boreal forest by creating a dynamic landscape mosaic with patches of different size, shape, age, and species composition. This mosaic and diversity at the stand and landscape levels are largely a product of the variability in wildfire (e.g., intensity, severity, size, shape, and frequency) and site characteristics (e.g., preburn vegetation and moisture regime). Since fuel, weather, and topography influence wildfire behavior, wildfire in essence behaves as multiple, dynamic treatments resulting in spatially variable effects. This variability occurs both within wildfires and between wildfires. Parisien et al. (2011) attribute spatial variability of area burned in the boreal forest to complex and nonstationary wildfire−environment relationships.

Bell (1889) observed a connection between boreal plant and animal species, and frequent wildfires. This boreal forest wildfire regime shapes the plant traits, favoring cohabitation with fire as a constant evolutionary pressure (Keeley et al., 2011). Pioneer tree species such as aspen, white birch, jack pine, and lodgepole pine usually recolonize burned sites in the boreal forest. These fast growing, shade-intolerant species are well adapted to landscapes with frequent, large, high-intensity wildfires. Aspen, balsam poplar, and birch use two strategies to establish quickly in burned areas. Adjacent unburned trees produce large amounts of light-weight seed capable of being carried long distances by wind to the burned sites. These seeds require a mineral soil seedbed for successful germination. When burned, aspen, balsam poplar, and birch trees are also capable of regenerating by sprouting from their roots and stumps. Lodgepole pine and jack pine have serotinous cones and only regenerate from seeds. The high temperatures generated from moderate- to high-intensity wildfires melt the resin binding the cone scales, thereby resulting in a seed rain after the passage of the flame front. The post-fire environment promotes seedling establishment and starts the survival cycle of regeneration.

Black spruce is another boreal species well adapted to coexist with wildfire (Viereck, 1983). It has semiserotinous cones, and high seed production at an early age. It can also reproduce by a vegetative process called layering. As the lower branches of black spruce trees grow downward and become covered with surface vegetation, they form independent root networks and new parent trees. The dense foliage, low crown base height, and overall complex structure of the stand (multiage and height) create ladder fuels for wildfire to quickly move from the surface to the crown. Black spruce stands are therefore highly flammable and difficult to contain when they burn. These stands are typically perpetuated by large, high-intensity wildfires.

Pre-fire stand characteristics (species composition and density) and wildfire severity influence postfire succession (Schimmel and Grandstrom, 1996; Lecomte et al., 2006). Regeneration density after the Richardson Fire in 2011 in northeast Alberta was the lowest in young jack pine stands that were severely burned, suggesting predicted climate changes may alter postfire succession (Pinno et al., 2013). Brown and Johnstone (2012) reported similarly reduced seed availability in black spruce stands due to repeat fires. Altered nutrient sinks due to increased fire severity and depth of organic consumption may also initiate alternate successional pathways with more deciduous trees postfire (Shenoy et al., 2013).

Boreal forest species' postwildfire plant regeneration strategies ensure next generation growth at the expense of the parent plants. In Canada's boreal, wildfire-dependent ecosystems, plant survival strategies enable vegetation to survive and thrive. However, these same conditions make it difficult for people and their communities to coexist in these wildfire-dependent ecosystems. Stocks (1991) suggested it is neither economically feasible nor ecologically desirable to eliminate all wildfires at all times. Balancing the positive and negative effects of wildfires is a challenge for wildfire management agencies.

3.2 WILDFIRE CAUSES AND IMPACTS

Aboriginal peoples in Canada (First Nations, Métis, and Inuit) have a long history of using fire. For example, it is estimated that the practice of traditional burning in northern Alberta dates back approximately 8,500 years (Lewis, 1982). Fire was critical for hunting, and was used by Aboriginal peoples to open up hunting trails, create wildlife habitat, influence game-herd migration, increase berry production, and promote the growth of new trees for firewood (Pyne, 2007b). It also opened up and cleaned areas used for camp sites and settlements, and reduced the risk of wildfire to encampments.

As European settlement and resource development in Canada moved westward, so did the occurrence of human-caused wildfires (Pyne, 2007b). In 1889, Bell estimated burned areas accounted for one-third of the total area of the boreal forest, areas of small postwildfire regeneration accounted for another one-third, with mature forests making up the remaining one-third. Bell's observational estimate of the burn rate suggests a period of high wildfire activity in the boreal forest before European settlement. Although Bell considered lightning to be the most common cause of wildfires in the late 1880s, he expressed concern about the increasing number of human-caused wildfires. The reasons he ascertained were the availability of "lucifer matches," and more people (missionaries, surveyors, explorers, prospectors, loggers, and road and railway construction crews) traveling through the forest lighting fires for smudges and campfires.

The most destructive wildfires in terms of injury, loss of life, and structures in Canadian history were eight human-caused events between 1825 and 1938. The most significant loss of life occurred during the 1916 Matheson fire in Ontario. A conservative estimate puts the loss at 223 fatalities (Alexander, 2010). These wildfires occurred mainly in late fall or early spring with preceding periods of drought. Strong winds, fast wildfire spread, and the occurrence of multiple ignitions contributed to the large area burned and extensive destruction of life and property.

Recent causes of wildfire are about equally divided between human and lightning at the national level, but high variability occurs at the regional level. For example, 86 percent of wildfires in the North West Territories are lightning caused, whereas 97 percent of wildfires in Nova Scotia are human caused. Human-caused wildfires continue to challenge wildfire management agencies since many of these wildfires threaten high values at risk.

Although a large area is burned each year in Canada, no civilians have been killed by being entrapped or overrun by wildfire, and few homes have been destroyed since the 1938 Dance Township Fire in Ontario (Alexander, 2010). Beverly and Bothwell (2011), for example, estimate that on average 18 homes were destroyed each year between 1980 and 2007. Two years, however, stand out as exceptions to this low level of loss: 2003 and 2011. The extremely hot and dry summer of 2003 resulted in >2,500 wildfires, with a significant number of wildfires encroaching onto communities. In British Columbia, >338 homes and many businesses were destroyed or damaged, and >45,000 people were evacuated (Filmon, 2004). Communities directly affected by these wildfires included cities (e.g., Kelowna, population of 106,707 in 2007), towns and villages of a few thousand residents, and small communities with populations of <1,000. Two air tanker crew members and a helicopter pilot lost their lives. Estimated costs of the British Columbia wildfires included approximately $200 million in insurance payouts, >$4.5 million for assistance and community support by the Red Cross, loss of 200 direct jobs when a saw mill was destroyed, clean-up costs of $6 million in Kelowna, and $13.5 million for the reconstruction of a historic site (Peter et al., 2006).

In May 2011, the Flat Top Complex wildfires entered the Town of Slave Lake, Alberta, destroying 344 family dwellings, six apartment buildings, three churches, 10 businesses, and the government center (the municipal library, town administrative offices, and regional provincial government offices). Fifty-six homes were also destroyed in the nearby communities of Canyon

Creek, Widewater, and Poplar Estates. These wildfires resulted in the evacuation of 10,000 residents in the Lesser Slave Lake Region for two weeks (Flat Top Complex Wildfire Review Committee, 2012). Claimed insurable losses were estimated at $742 million (Insurance Bureau of Canada, 2012), wildfire-suppression costs at $16 million, and recovery costs at $289 million, bringing the total costs to >$1 billion (Flat Top Complex Wildfire Review Committee, 2012; KPMG, 2012). There was also evidence of social impacts as residents reevaluated life goals and priorities and experienced high levels of stress; family routines were disrupted; higher rates of vandalism, disturbing the peace, and mental health issues were reported by police; and social relationships were transformed (KPMG, 2012; Pujadas Botey and Kulig, 2013).

The effects of smoke from wildfires are a growing concern. Wildfire smoke can reach high concentrations and have adverse health impacts on communities near wildfires and those living a considerable distance away (Johnston and Bowman, 2013). Infants, the elderly, pregnant women, and people with respiratory and cardiovascular conditions are particularly sensitive (Johnston and Bowman, 2013). In the event that smoke from a nearby wildfire affects a community, the smoke-sensitive populations are evacuated first—sometimes by air in the case of remote communities. Smoke can also have impacts a considerable distance away from the wildfire. In 2013, about 500,000 customers were without power in the Montreal urban area, disrupting business and causing havoc with public transit, after smoke from wildfires in remote areas of northern Quebec affected hydroelectric power transmission (CTV Montreal 2013). During the wildfire season, smoke forecasting occurs for all of Canada south of the Arctic Ocean. Forecasts provide information to predict air quality and issue health advisories.

Infrastructure impacts of wildfires have also been significant (Tymstra, 2013). For example, the 2011 wildfires in Alberta caused facility, pipeline, maintenance, and new project construction shutdowns resulting in a decrease in oil and gas extraction of 3.6 percent in the second quarter (Statistics Canada, 2011). This contributed to a decrease in Canada's Real Gross Domestic Product of 0.1 percent for the same period (Statistics Canada, 2011).

3.2.1 At-Risk Communities

Fortunately, Canada has low population densities in areas with the highest wildfire activity. This has helped prevent wildfire losses (Beverly and Bothwell, 2011). A majority of Canadians live in urban areas removed from areas of high wildfire activity. Populations in these areas continue to grow at a fast rate. In contrast, the total population in rural regions has shown a modest increase, with most of this increase in communities adjacent to urban centers (Bollman and Clemenson, 2008). There are many communities and rural populations, however, adjacent to or scattered throughout Canada's forested regions and at potential risk from wildfire (Figure 3.2.). Some of these communities, particularly those with natural resource dependent economies (e.g., Fort McMurray, Alberta) and those with high natural amenity values (e.g., Kelowna, British Columbia), are experiencing rapid population growth and expansion of their wildland–urban interface.

Many Aboriginal communities in Canada face high risk from wildfire due to their location in forests that burn frequently (Wotton and Stocks, 2006). About one third of wildfire evacuations in Canada involve Aboriginal communities (Beverly and Bothwell, 2011). Aboriginal peoples are among the fastest growing populations in Canada. Statistics Canada (2005) predicted populations living on reserves and settlements will increase by 40 percent between 2001 and 2017.

Increasing industrial development and recreational use pose risks of human-caused wildfires, increases demand on firefighting resources, and heightens the need for risk mitigation. For example, in a review of wildfire starts in Alberta's forest protection zone over a 10-year period (1996–2005), industry (e.g., oil and gas, forest industry, railroad, and power lines) was found to be the third highest human-caused ignition source (Canadian Association of Petroleum Producers, 2008). Flaring from gas wells, brush burning, and the use of all-terrain vehicles account for about three quarters of oil and gas industry wildfire starts. The oil and gas industry has responded by developing best practices for wildfire prevention and a guidebook for assessing the threat of wildfire on dispositions and mitigating the risk and liability to the industry (Alberta Sustainable Resource Development, nd). Increased development of recreational cottages in areas with high natural amenity values (e.g., lakeshore or mountains) is widespread across Canada (Halseth, 2004). In the past, recreational cottages have served as summer, weekend getaways for residents within commuting distance of urban centers. Recently, however, the transformation of these small seasonal cottage areas into high value, year-round resort developments such as the mountain resort towns of Canmore, Alberta, and Whistler, British Columbia, are creating new challenges for wildfire management. Absentee owners who place a high value on forest cover may be difficult to reach with educational messages and to engage in risk-reduction initiatives.

3.2.2 Land Use and Institutional Arrangements

Most (93 percent) of Canada's forest and wooded lands are under public ownership (Natural Resources Canada, 2012), with a maximum of 99 percent in Newfoundland and Labrador and a minimum of 8 percent in Prince Edward Island. The 10 provincial and three territorial governments are the primary forest landowners in Canada (77 percent of forest and wooded land) and forest and wildfire management is the responsibility of the respective jurisdiction. The 7 percent of forested land in private ownership belongs to >450,000 landowners.

The provincial and territorial wildfire management agencies undertake several activities related to prevention, preparedness, and suppression. Prevention usually includes education aimed at the public, industry, and other forest users to reduce the number of human-caused wildfires. Initiating awareness campaigns, issuing fire permits for burning, implementing fire bans and forest closures (through legislative tools) during times of high wildfire risk are common tools used to help prevent ignitions. Prevention also includes actions to reduce risk before wildfire occurs. Some wildfire management agencies integrate mitigation actions such as thinning, logging, and prescribed fire into forest-management planning to reduce the risk of high-intensity wildfires at the landscape level. Several wildfire-management agencies also provide assistance to communities in developing community protection plans, providing expertise and resources, and in some cases, financial grants for completing mitigation in the community.

Wildfire management agencies implement preparedness actions leading to a state of readiness to cope with potential wildfires. During the wildfire season, agencies monitor the fire danger and preposition suppression resources to areas where the forecasted threat from wildfire (hazard, fire occurrence, and values at risk) is the greatest. Preparedness also includes early fire detection through aerial surveillance, staffed lookouts, and forest patrols.

When a wildfire is detected and reported, wildfire management agencies initiate an appropriate suppression response based on policy. Resources (initial attack and sustained action crews, and aircraft) are typically dispatched to achieve a specific initial attack size objective. Crews consist of paid, full-time, and seasonal-trained fire fighters; contract resources (including fire fighters, equipment, and aircraft) are used if necessary.

The federal government is responsible for wildfire management in a relatively small percentage (16 percent) of the forests in Canada; this area comprises Aboriginal communities, military bases, and national parks. The federal government maintains agreements with provincial wildfire agencies to provide fire suppression resources on Aboriginal settlements and national defence lands. The national parks agency, Parks Canada, has its own wildfire management capability including planning, public education, and wildfire suppression and prescribed burning (i.e., a controlled burn under prescribed conditions). Parks Canada pioneered the return of fire to protected areas in Canada. Banff National Park began prescribed burning in 1983 using ecological classification and fire history data to assign fire cycles to burn units within the park (White et al., 2011). The program transitioned to focus on building containment areas using mechanical thinning combined with pre-scribed burning, to mitigate the threat of high intensity fires escaping (White et al., 2011). To support ecological restoration and the implementation of a new Banff National Park Management Plan, the prescribed burn program began taking an ecocultural focus in 1997 (White et al., 2011).

The federal forest agency, the Canadian Forest Service (CFS), does not manage forested lands and therefore has no direct role in operational wildfire management. The CFS instead is a science-based policy organization within Natural Resources Canada. The primary role of the CFS in relation to wildfire is to increase knowledge about wildfires in Canada through research that aims to strengthen the ability of wildfire management agencies to predict and manage wildfire risks and benefits.

Local governments play key roles in wildfire management. Wildfires on private lands in or near communities are generally suppressed by local fire departments and are not included in provincial wildfire management agency or national reporting. Anecdotal evidence, however, suggests that a significant number of these wildfires occur each year. Many rural community fire departments rely on volunteer fire fighters, and they often supplement their wildfire resources through mutual aid agreements with neighboring jurisdictions and provincial and territorial wildfire agencies and local industries. Although the terms of the agreements vary, they generally permit the sharing of personnel and equipment during wildfires, and wildfire information and cooperative mitigation and preparedness activities (e.g., crosstraining). Local governments have the responsibility for planning and development of lands within their boundaries, and many are also actively involved in their own wildfire mitigation, prevention and education programs. Studies in Alberta and British Columbia (Harris et al., 2011; McGee and Labossiere, 2013) have found that in addition to the mandatory emergency plans and infrastructure measures (access to water, adequate road widths), many local governments are implementing wildfire mitigation measures. Vegetation management, including clearing, pruning and fireguards are also being implemented. Education is a popular tool being used for both internal and external audiences. Although labor intensive, some local governments are completing wildfire hazard assessments on homeowners' properties. Land use planning measures are being used to mitigate wildfire risks, including, in some cases a requirement for homeowners to implement mitigation measures on their property. In most cases however, wildfire mitigation by homeowners remains voluntary.

3.3 WILDFIRE MANAGEMENT

Wildfire suppression has been the dominant approach to wildfire in Canada for >100 years. Suppression strategies vary between provinces, territories, and national parks due to different landscapes, fire load, forest ownership, land management objectives, and values at risk. Most wildfire management agencies, however, use a zone system whereby lands are classified as full suppression, modified suppression or no suppression, based primarily on the values at risk. Agencies not using zonation apply a risk assessment approach that recognizes the spatial and temporal changes in risk. This approach assesses risk on a wildfire-by-wildfire basis to determine the appropriate response to help achieve land management objectives. Appropriate response may include the use of prescribed wildfire (i.e., allowing wildfire to burn under prescribed conditions). Wildfire management agencies also conduct prescribed burns to return wildfire to wildfire dependent ecosystems or to meet specific resource-management objectives. Fire managers conduct detailed planning before prescribed burns and prescribed wildfire are carried out. Regardless of variation in strategies, fire fighting in Canada is highly effective (Cummings, 2005; Martell and Sun, 2008). However, wildfire suppression efforts face challenges from the loss of valuable expertise because of staff retirements, recruitment and retention of young fire fighters, and aircraft and equipment modernization (Born and Stocks, 2006).

Wildfire suppression agencies share firefighting resources to help reduce costs and improve efficiency and effectiveness. The Canadian Interagency Fire Center (CIFFC) has coordinated wildfire suppression resource sharing between provincial, territorial and federal agencies since 1982. The Canadian Interagency Mutual Aid and Resource Sharing Agreement facilitates the sharing of equipment, personnel, and aircraft for wildfire suppression among wildfire management agencies. CIFFC develops and promotes a national approach and standards for aviation assets, personnel, equipment, resource sharing, and geospatial tools and applications and promotes collaboration on wildfire science and technology, knowledge exchange, wildfire prevention, and wildfire meteorological issues. CIFFC also coordinates the sharing of resources with other countries. Formal agreements between Canada and the United States allow rapid deployment across the international boundary and Canada provides resource sharing with other countries as needed.

3.3.1 Wildfire Mitigation

Despite the historical success of wildfire suppression, increasing impacts of wildfire on people and property, predicted increases in fire occurrence and intensity that may overwhelm fire suppression capacity (Podur and Wotton, 2010), and recognition that wildfire is essential to healthy ecosystems suggest that a new approach is needed for wildfire management in Canada (Canadian Wildland Fire Strategy Assistant Deputy Ministers Task Group, 2005). In 2005, the Canadian Council of Forest Ministers developed a shared vision for wildfire management among the provincial, territorial, and federal governments. The Canadian Wildland Fire Strategy (CWFS) calls for more "comprehensive risk management approaches, including an appropriate mix of mitigation, preparedness, response, and recovery" (Canadian Wildland Fire Strategy Assistant Deputy Ministers Task Group, 2005). Suppression remains the primary strategy, with most provinces and territories dedicating only a small proportion of wildfire management resources to mitigation programs.

The Partners-in-Protection Association (PiP) is a nonprofit organization whose activities complement those of provincial, territorial, and municipal agencies by creating and distributing educational materials, developing information forums, and encouraging community-based initiatives to reduce wildfire risk and enhance safety to communities. Membership in PiP includes a diverse group of stakeholders with an interest in community protection, including national, provincial and municipal associations, government departments responsible for emergency services, forest and park management, land-use planning, and private business and industry. PiP developed the first comprehensive technical manual on wildfire mitigation in Canada (Partners-in-Protection, 1999). The manual, now in its second edition, represents an unofficial Canadian standard for wildfire mitigation and preparedness, and has been distributed across Canada and internationally. In the spring of 2012, PiP launched FireSmart Canada® with the goal of facilitating shared responsibility between governments, communities, and citizens in reducing wildfire risk to communities.

Effective mitigation requires both public support for government action and engagement of private property owners in mitigating the risk to their own properties. Studies have shown a high level of public support for mitigation efforts by wildfire management agencies or local governments in western Canada. For example, Flanagan (2008) found that among six high risk communities in Alberta, there was strong support for education, free hazard assessments completed by government agency personnel on homeowners' properties, and for by-laws requiring homeowners to use fire resistant building materials (Table 3.1.). More than half of survey respondents in Flanagan's

TABLE 3.1 Percent of Homeowners in Alberta Who Support Wildfire Mitigation Policies and Fuels Management Options

	n	Percent
Wildfire Mitigation Policies		
Educate homeowners about ways to reduce their wildfire risk	1,250	90.2
Free residential wildfire hazard assessments	1,241	80.5
Bylaws requiring new houses to use fire-resistant building materials	1,248	77.4
Bylaws requiring homeowners to remove vegetation close to their house	1,252	57.5
Restrict building in high-risk areas	1,248	55.7
Neighborhood work bees	1,247	47.0
Fuel Management Options		
Fireguards around communities	1,253	79.2
Prescribed burning	1,252	73.0
Thinning trees	1,253	70.9

Source: From Flanagan, 2008.

study supported by-laws requiring residents to remove vegetation on their property and building restrictions in high-risk areas. Neighborhood work bees to help reduce the local wildfire risk were supported by less than half of respondents. Support for fuel modification (establishing fireguards around communities, using prescribed fire and thinning vegetation) at the wider community and landscape level also received strong support (>70 percent). Although there appears to be general support for many wildfire mitigation measures by governments, it is important to note that public support will depend on how these mitigation measures are carried out at the local level. For example, taking residents' values and concerns into account and engaging residents throughout plan development and implementation will likely enhance the likelihood of support (Christianson et al., 2012; McGee, 2011).

Many Alberta homeowners have completed mitigation actions on their property (e.g., cutting and watering grass; removing shrubs, trees, and fallen branches close to the house; thinning shrubs and trees; removing tree branches close to the ground; removing needles, leaves, and branches from roof and gutters), but some mitigation measures (landscaping with fire-resistant materials and vegetation; screening house vents, gutters, and under eaves with mesh; screening or enclosing under decks and porches, and installing stucco, metal, brick, or other fire-resistant exterior siding on the house) are less popular. Homeowners likely to implement mitigation measures are older, perceive the wildfire threat to be significant enough to undertake mitigation measures, have the financial resources to implement mitigation measures, view the impacts from wildfires to be controllable, and place a high priority on the completion of mitigation measures (Flanagan, 2008; McFarlane et al., 2011).

There are also examples in Canada where residents have become actively engaged in wildfire mitigation in and around their neighborhoods. One example from the province of Alberta is the FireSmart-ForestWise program in Jasper National Park (Westhaver, 2006) where cottage owners around Lake Edith are active participants in work bees to assist Parks Canada with vegetation management around their neighborhood (McGee, 2011).

3.3.2 Community Response to Wildfire

In Canada, when public safety is threatened by wildfire (e.g., when a wildfire comes close to a community, escape routes are in danger of being cut-off or there are concerns related to smoke), evacuation is the approach supported by governments to protect public safety. Between 1980 and 2007, there were 547 wildfire evacuations that resulted in the evacuation of 209,121 people (an average of about 7,500 people each year) (Beverly and Bothwell, 2011). Most evacuations have occurred in the boreal forest; however, evacuations have also occurred outside the boreal where there are more communities and people at risk (Figure 3.3).

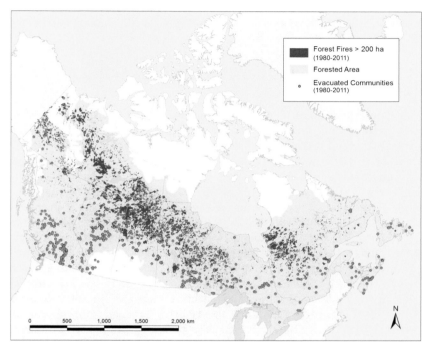

FIGURE 3.3 Large wildfires (>200 ha) and evacuated communities in Canada. © 2013 BL McFarlane. *(Forested area is derived from Rowe (1972) and large fires are from the National Fire Database 1980–2011.)* Evacuation data are from the Natural Resources Canada wildfire evacuation data base.

During a wildfire, the agency responsible for managing the wildfire provides information to community officials about the potential threat to public safety. These officials may then issue an evacuation order. Evacuations may include those at high risk from smoke (including infants, the elderly, pregnant women, and people with respiratory and cardiovascular conditions—stage 1 evacuation), or all community members (stage 2 evacuation). Stage 1 evacuations may occur due to heavy smoke even if the wildfire itself does not pose a direct threat to the community. Depending on wildfire behavior, wind direction, and other factors, some locations may be evacuated more than once during a wildfire. Once an evacuation order is issued, multiple agencies are involved in helping to carry out the evacuation, including local government, provincial and federal governments, police (local, provincial, and national) and nongovernment organizations (e.g., the Red Cross). In cases where remote communities do not have road access, residents are evacuated by air.

When an evacuation order is issued, it is considered mandatory and citizens are obligated to comply. However, some citizens may stay behind. For example, in the Mt McLean fire in the Squamish Lillooet District of British

Columbia, 100 to as many as 450 people were estimated to have stayed behind (James, 2009; The Vancouver Sun, 2009).

3.4 ECOLOGICAL RESTORATION AND COMMUNITY RECOVERY

It is common practice across Canada to evaluate opportunities to harvest trees damaged by wildfire, on the basis that if not salvaged, the resource is lost, and a potential fuel buildup and increased fire hazard may occur. There is increasing concern, however, about the ecological effects of post-fire salvage harvesting, and in particular, the cumulative impacts (Purdon et al., 2004; Macdonald, 2007). The application of science-based guidelines for managing newly burned forests is important to help mitigate short- and long-term impacts. British Columbia has adopted an approach similar to the Burn Area Emergency Response program in the United States where teams are dispatched to address immediate and significant threats from post-fire impacts (e.g., potential impact on drinking water supply due to erosion and sediment flow). In British Columbia, analogous teams are deployed to selected wildfires to identify and address post-fire slope stability impacts. The 2003 Kelowna wildfire highlighted the importance of stabilizing critical slopes to control erosion (Jordan et al., 2009). External fire reviews are typically completed after disastrous fire seasons, and some such as the Firestorm 2003 Provincial Review Report address the need to restore areas impacted by wildfires to mitigate post-fire floods and landslides (Filmon, 2004).

The 2011 Flat Top Complex wildfires in Alberta and the 2003 wildfires in British Columbia have shown that wildfires can have major financial implications for individuals, communities, businesses, and governments in Canada. After a wildfire, the insurance industry provides coverage for damage to homes and businesses when owners have adequate insurance coverage. Purchasing insurance, however, is voluntary, and not all homeowners or businesses are insured or they often underinsure their assets (KPMG, 2012). With the exception of the 2003 and 2011 wildfire seasons, insurable losses in Canada have been moderate: <1 percent of overall losses (Kovacs, 2001). Provincial and territorial governments may also step in to provide financial assistance to homeowners and businesses after wildfires (KPMG, 2012). Government relief, however, is not generally provided to property owners if private insurance is available (Kovacs and Kunreuther, 2001). Organizations such as the Red Cross and local community-based groups may also provide support to communities during the recovery process by collecting financial and physical donations and providing volunteers to assist with recovery efforts.

The Government of Canada provides financial assistance to provincial and territorial governments, and in turn to local governments and communities for recovery and relief in the wake of a large-scale disaster. The federal government only becomes involved, however, at the request of provincial or territorial governments or if the disaster is national in scope (Public Safety Canada, 2013). There are several federal programs for disaster relief including the Disaster Financial Assistance Arrangements (DFAA), which provides funds to provincial and territorial governments to help them meet the costs arising from physical damage caused by major disasters that would otherwise impose an excessive burden on their economies and citizens. Through the DFAA, assistance is paid to the province or territory—not directly to affected individuals, small businesses, or communities.

3.5 CONCLUSIONS

Wildfire is an integral component in shaping Canada's forests. Most years, the fire-dependent forest ecosystems in Canada are significantly impacted by fire. This however has not translated into commensurate levels of home losses or damages to communities. Typically, several communities are threatened each year and evacuated, but since 1938, only two years experienced major losses. Effective wildfire suppression and low population densities in areas with high wildfire activity contribute to this low level of loss. However, as populations grow and industrial development expands into forested areas and climate change alters fire regimes, the risk to communities and infrastructure will likely increase. As wildfire-suppression costs escalate, wildfire management agencies are putting more emphasis on mitigating the risk before wildfires occur. The CWFS outlined a vision in 2005 for the future of wildfire management in Canada. While progress has been made, further efforts are needed to improve collaboration and cooperation across the country, and to increase support for wildfire mitigation efforts within provinces, local governments, communities, and by homeowners.

ACKNOWLEDGMENTS

The authors thank John Little and Peter Englefield, Natural Resources Canada for supplying the maps and Rob McAlpine from the Ontario Ministry of Natural Resources, Hugh Boyd and Bruce Mayer of Alberta Environment and Sustainable Resource Development, and Bruce Macnab from Natural Resources Canada, Canadian Forest Service for providing comments on earlier drafts.

REFERENCES

Alberta Sustainable Resource Development, nd. FireSmart® Guidebook for the Oil and Gas Industry. Available from: http://srd.alberta.ca/Wildfire/FireSmart/documents/FireSmart-Guidebook-OilAndGasIndustry-2008.pdf (accessed 26.11.13.)

Alexander, M.E., 2010. 'Lest we forget': Canada's major wildland fire disasters of the past, 1825–1938. In: Proceedings of the 3rd Fire Behavior and Fuels Conference, Oct. 25–29, 2010, Spokane, Washington. International Association of Wildland Fire, Birmingham, Alabama, pp. 1–21.

Alexander, M.E., Heathcott, M.J., Schwanke, R.L., 2013. Fire Behaviour Case Study of Two Early Winter Grass Fires in Southern Alberta, 27 November 2011. Partners in Protection Association, Edmonton, Alberta.

Anielski, M., Wilson, S., 2009. The Real Wealth of the Mackenzie Region: Assessing the Natural Capital Values of a Northern Boreal Ecosystem. Canadian Boreal Institute, Ottawa, ON.

Bell, R., 1889. Forest fires in northern Canada. Proc. Am. Forestry Congress 7, 50−55.

Beverly, J.L., Bothwell, P., 2011. Wildfire evacuations in Canada 1980−2007. Nat. Hazards 59, 571−596. http://dx.doi.org/10.1007/s11069-011-9777-9.

Bollman, R.D., Clemenson, H.A., 2008. Structure and change in Canada's rural demography: an update to 2006. Rur. Small Town Can. Anal. Bull. 7 (7). Statistics Canada, Ottawa, ON. Available from: http://www.statcan.gc.ca/pub/21-006-x/21-006-x2007007-eng.pdf (accessed 26.11.13.).

Born, W., Stocks, B.J., 2006. Canadian fire management infrastructure. In: Hirsch, K.G., Fuglem, P. (Eds.), Canadian Wildland Fire Strategy: Background Syntheses, Analyses, and Perspectives, pp. 57−71. Available from: http://cfs.nrcan.gc.ca/pubwarehouse/pdfs/26529.pdf (accessed 26.11.13.).

Boulanger, Y., Gauthier, S., Burton, P.J., Vaillancourt, 2012. An alternative fire regime zonation for Canada. Int. J. Wildland Fire 21 (8), 1052−1064.

Brown, C., Johnstone, J.F., 2012. Once burned, twice shy: repeat fires reduce seed availability and alter substrate constraints on *Picea mariana* regeneration. Forest Ecol. Manag. 266, 34−41. http://dx.doi.org/10.1016/j.foreco.2011.11.006.

Canadian Association of Petroleum Producers, 2008. Best Management Practices. Wildfire Prevention. Canadian Association of Petroleum Producers, Calgary, AB. Available from: http://membernet.capp.ca/raw.asp?x=1&dt=NTV&dn=132380 (accessed 26.11.13.).

Canadian Council of Forest Ministers, 2005. Canadian wildland fire strategy: a vision for an innovative approach to managing the risks. A report to the Canadian Council of Forest Ministers, prepared by the Canadian Wildland Fire Strategy Assistant Deputy Ministers Task Group.

Canadian Wildland Fire Strategy Assistant Deputy Ministers Task Group, 2005. Canadian Wildland Fire Strategy: A Vision for an Innovative and Integrated Approach to Managing the Risks. A report to the Canadian Council of Forest Ministers. Available from: http://cfs.nrcan.gc.ca/pubwarehouse/pdfs/26218.pdf (accessed 26.11.13).

Christianson, A., McGee, T.K., L'Hirondelle, L., 2012. Community support for wildfire mitigation at Peavine Métis Settlement, Alberta, Canada. Environ. Hazards 11 (3), 177−193. http://dx.doi.org/10.1080/17477891.2011.649710.

Commissioner of the Environment and Sustainable Development, 2013. 2013 Fall Report of the Commissioner of the Environment and Sustainable Development (Chapter 1—Backgrounder on Biological Diversity). Office of the Auditor General of Canada, Ottawa, 32 p.

Cumming, S.G., 2005. Effective fire suppression in boreal forests. Can. J. Forest Res. 35, 772−786.

Ecological Stratification Working Group, 1995. A National Ecological Framework for Canada. Agriculture and AgriFood Canada, Research Branch, Centre for Land and Biological Resources Research and Environment Canada, State of the Environment Directorate, Ecozone Analysis Branch, Ottawa/Hull. Report and national map at 1:7 500 000 scale.

Filmon, G., 2004. Firestorm 2003 Provincial Review: A Report to the Province of British Columbia. Available from: http://bcwildfire.ca/history/reportsandreviews/2003/firestormreport.pdf (accessed 26.11.13.).

Flanagan, H., 2008. Residential Wildfire Mitigation Preferences. (Unpublished Master of Arts thesis), University of Alberta, Edmonton, Canada.

Flannigan, M.D., Krawchuk, M.D., de Groot, W.J., Wotton, B.M., Gowman, L.M., 2009. Implications of changing climate for global wildland fire. Int. J. Wildland Fire 18, 483—507.

Flannigan, M.D., Cantin, A.S., de Groot, W.J., Wotton, B.M., Newbery, A., Gowman, L.M., 2013. Global wildland fire severity in the 21st century. Forest Ecol. Manag. 294, 54—61.

Flat Top Complex Wildfire Review Committee, 2012. Flat Top Complex. Alberta Environment and Sustainable Resource Development Pub. No. T/272, Edmonton, AB. Available from: http://srd.alberta.ca/Wildfire/WildfirePreventionEnforcement/WildfireReviews/documents/FlatTopComplex-WildfireReviewCommittee-May18-2012.pdf (accessed 26.11.13.).

de Groot, W.J., Flannigan, M.D., Cantin, A.S., 2013. Climate change impacts on future boreal fire regimes. Forest Ecol. Manag. 294, 35—44.

Halseth, G., 2004. The 'cottage' privilege: increasingly elite landscapes of second homes, in Canada. In: Hall, C.M., Müller, D. (Eds.), Tourism, Mobility and Second Homes: Between Elite Landscape and Common Ground. Channel View Publications, Clevedon, UK, pp. 35—54.

Harris, L.M., McGee, T.K., McFarlane, B.L., 2011. Implementation of wildfire risk management by local governments in Alberta, Canada. J. Environ. Plan. Manag. 54 (4), 457—475. http://dx.doi.org/10.1080/09640568.2010.515881.

Insurance Bureau of Canada, 2012. Understanding Insurance. http://www.ibc.ca/en/otherdocs/Understanding%20Ins%20brochure_EN.pdf (accessed 08.12.13.).

James, A., August 12, 2009. Staying Behind, [Letter to the Editor]. Bridge River Lillooet News. Retrieved from: http://www.lillooetnews.net/article/20090812/LILLOOET0303/308129927/staying-behind#art_comments (accessed 08.12.13.).

Johnston, F., Bowman, D., 2013. Bushfire smoke: an exemplar of coupled human and natural systems. Geogr. Res. 52 (1), 45—54. http://dx.doi.org/10.1111/1745-5871.12028.

Jordan, P., Covert, A., Curran, M., 2009. Post-wildfire mass movement and erosion events in British Columbia, 2003—2009. In: Presentation to the Wildfire and Watershed Hydrology Workshop, Kelowna, June 3—4, 2009. http://www.for.gov.bc.ca/hfd/library/FIA/2010/FSP_Y103004h.pdf (accessed 08.12.13.).

Keeley, J.J., Pausas, J.G., Rundel, P.W., Bond, W.J., Bradstock, R.A., 2011. Fire as an evolutionary pressure shaping plant traits. Trends Plant Sci. 16 (8), 406—411. http://dx.doi.org/10.1016/j.tplants.2011.04.002.

Kovacs, P., 2001. Wildfire and Insurance. ICLR Research Paper Series—No. 11. Available from: http://www.iclr.org/images/Wildfires_and_insurance.pdf (accessed 20.11.13.).

Kovacs, P., Kunreuther, H., 2001. Managing catastrophic risk: lessons from Canada. In: Paper Presented at the ICLR/IBC Earthquake Conference, March 23, 2011. Simon Fraser University, Vancouver, BC. Available from: http://www.iclr.org/images/Managing_Catastrophic_Risk.pdf (accessed 20.11.13.).

KPMG, 2012. Lesser Slave Lake Regional Urban Interface Wildfire—Lessons Learned. Final Report. Available from: http://www.aema.alberta.ca/documents/0426-Lessons-Learned-Final-Report.pdf (accessed 26.11.13.).

Krezek-Hanes, C.C., Ahern, F.J., Cantin, A., Flannigan, M.D., 2011. Trends in Large Fires in Canada, 1959—2007. Canadian Biodiversity: Ecosystem Status and Trends 2010, Technical Thematic Report No. 6. Canadian Council of Forest Ministers, Ottawa, ON. http://www.biodivcanada.ca/default.asp?lang=En&n=137E1147-0.

Lecomte, N., Simard, M., Bergeron, Y., 2006. Effects of fire severity and initial tree composition on stand structural development in the coniferous boreal forest of northwestern Quebec, Canada. Ecoscience 13 (2), 152—163.

Lewis, H.T., 1982. A Time for Burning, 17. Boreal Institute for Northern Studies, University of Alberta.

Liu, Y., Stanturf, J., Goodrick, S., 2010. Trends in global wildfire potential in a changing climate. Forest Ecol. Manag. 259, 685−697.

Montreal, C.T.V., 2013. Forest Fires Cause Montreal Area Power Outages. Available from: http://montreal.ctvnews.ca/forest-fires-cause-montreal-area-power-outages-1.1353895 (accessed 5.11.13.).

Macdonald, S.E., 2007. Effects of partial post-fire salvage harvesting on vegetation communities in the boreal mixedwood forest region of northeastern Alberta, Canada. Forest Ecol. Manag. 239, 21−31.

Mack, M.C., Bret-Harte, M.S., Hollingsworth, T.N., Jandt, R.R., Schuur, E.A.G., Shaver, G.R., Verbyla, D.L., 2011. Carbon loss from an unprecedented tundra wildfire. Nature 475, 489−492.

Martell, D.L., Sun, H., 2008. The impact of forest fire suppression, vegetation and weather on burned area in Ontario. Can. J. Forest Res. 38 (6), 1547−1563.

McFarlane, B.L., McGee, T.K., Faulkner, H., 2011. Complexity of homeowner wildfire risk mitigation: an integration of hazard theories. Int. J. Wildland Fire 20 (8), 921−931. http://dx.doi.org/10.1071/WF10096.

McGee, T.K., 2011. Public engagement in neighbourhood level wildfire mitigation and preparedness: case studies from Canada, the US and Australia. J. Environ. Manag. 92 (10), 2524−2532. http://dx.doi.org/10.1016/j.jenvman.2011.05.017.

McGee, T.K., Labossiere, L., 2013. Wildfire mitigation by local governments in Alberta and British Columbia, Canada. In: Presented at International Research Committee on Disasters Researchers' Meeting—July 17, 2013. Broomfield, Colorado.

National Forest Inventory, 2006. Terrestrial Ecozone Standard Reports. Canadian Council of Forest Ministers. https://nfi.nfis.org/publications/standard_reports/NFI_T4_FOR_AREA_en. html (accessed 08.12.13.).

Natural Resources Canada, 2012. The State of Canada's Forests: Annual Report 2012. Natural Resources Canada, Ottawa, ON. Available from: http://cfs.nrcan.gc.ca/pubwarehouse/pdfs/34055.pdf (accessed 26.11.13.).

Parisien, M.-A., Parks, S.A., Krawchuk, M.A., Flannigan, M.D., Bowman, L.M., Moritz, M.A., 2011. Scale-dependent controls on the area burned in the boreal forest of Canada. Ecol. Appl. 21 (3), 789−805.

Partners in Protection, 1999. FireSmart Protecting Your Community from Wildfire. Partners in Protection, Edmonton, AB.

Peter, B., Wang, S., Mogus, T., Wilson, B., 2006. Fire risk and population trends in Canada's wildland-urban interface. In: Hirsch, K.G., Fuglem, P. (Eds.), Canadian Wildland Fire Strategy: Background Syntheses, Analyses, and Perspectives, pp. 37−48. Available from: http://cfs.nrcan.gc.ca/pubwarehouse/pdfs/26529.pdf (accessed 26.11.13.).

Pinno, B.D., Errington, R.C., Thompson, D.K., 2013. Young jack pine and high severity fire combine to create potentially expansive areas of understocked forest. Forest Ecol. Manag. 310, 517−522.

Podur, J., Wotton, M., 2010. Will climate change overwhelm fire management capacity? Ecol. Modell. 221, 1301−1309.

Price, D.T., Alfaro, R.I., Brown, K.J., Flannigan, M.D., Fleming, R.A., Hogg, E.H., Girardin, M.P., Lakusta, T., Johnston, M., McKenney, D.W., Pedlar, J.H., Stratton, T., Sturrock, R.N., Thompson, I.D., Trofymow, J.A., Venier, L.A., 2013. Anticipating the consequences of climate change for Canada's boreal forest ecosystems. Environ. Rev. 21, 322−365. http://dx.doi.org/10.1139/er-2013-0042.

Public Safety Canada, 2013. Emergency Management. Recovery from Disasters. http://www.publicsafety.gc.ca/cnt/mrgnc-mngmnt/rcvr-dsstrs/index-eng.aspx (accessed 07.11.13.).

Pujadas Botey, A., Kulig, J.C., 2013. Family functioning following wildfires: recovery from the 2011 Slave Lake fires. J. Child Fam. Stud. http://dx.doi.org/10.1007/s10826-013-9802-6.

Purdon, M., Brais, S., Bergeron, Y., 2004. Initial response of understory vegetation to fire severity and salvage-logging in the southern boreal forest of Quebec. Appl. Veg. Sci. 7, 49−60.

PWSS, 2013. Fast Facts—Grasslands National Park Fire April 27−28, 2013. Prairie Wind & Silver Sage, Friends of Grasslands. http://pwss.org/PDFs/Fast%20Facts%20May%205%20for%20public.pdf (accessed 08.12.13.).

Pyne, S.J., 2007a. Burning border. Environ. Hist. 12 (Oct.), 950−965.

Pyne, S.J., 2007b. Awful Splendour: A Fire History of Canada. UBC Press, Vancouver BC.

Roberts, M.R., 2004. Response of the herbaceous layer to natural disturbance in North American forests. Can. J. Bot. 82, 1273−1283.

Rowe, J.S., 1972. Forest Regions of Canada. Publication # 1300. Canadian Forestry Services. Environment Canada, Ottawa.

Rowe, J.S., 1992. The ecosystem approach to forestland management. Forestry Chronicle 68 (2), 222−224.

Schimmel, J., Granström, A., 1996. Fire severity and vegetation response in the boreal Swedish forest. Ecology 778, 1436−1450.

Shenoy, A., Kielland, K., Jonstone, J.F., 2013. Effects of fire severity on plant nutrient uptake reinforce alternate pathways of succession in boreal forests. Plant Ecol. http://dx.doi.org/10.1007/S11258-013-0191-0.

Statistics Canada, 2005. Projections of the Aboriginal Populations, Canada, Provinces and Territories 2001−2017. Catalogue no. 91-547-XIE. Minister of Industry, Ottawa, ON. Available from: http://www.statcan.gc.ca/pub/91-547-x/91-547-x2005001-eng.pdf (accessed 26.11.13.).

Statistics Canada, Wednesday, August 31, 2011. Canadian economic accounts. The Daily. http://www.statcan.gc.ca/daily-quotidien/110831/dq110831a-eng.htm (accessed 08.12.13.).

Stocks, B.J., 1991. The extent and impact of forest fires in northern circumpolar countries. In: Levine, J.S. (Ed.), Global Biomass Burning: Atmospheric, Climatic, and Biospheric Implications. The MIT Press, Cambridge, Massachusetts, pp. 197−204.

Stocks, B.J., Mason, J.A., Todd, J.B., Bosch, E.M., Wotton, B.M., Amiro, B.D., Flannigan, M.D., Hirsch, K.G., Logan, K.A., Martell, D.L., Skinner, W.R., 2002. Large forest fires in Canada, 1959−1997. J. Geophys. Res. 108 (D1), 1−12.

Stonehouse, J.D., 2010. Ecoregions of Canada's prairie grasslands. In: Shorthouse, J.D., Floate, K.D. (Eds.), Ecology and Interactions in Grassland Habitats, Arthropods of Canadian Grasslands, vol. 1. Biological Survey of Canada, pp. 53−81.

The Globe and Mail, 2012. Massive Southern Alberta Grass Fires Force 3,500 to Leave Homes. September 10, 2013. http://www.theglobeandmail.com/news/national/massive-southern-alberta-grass-fires-force-3500-to-leave-homes/article4534808/ (accessed 08.12.13.).

The Pincher Creek Voice, January 6, 2012. Grassfires Ravage Areas of Southern Alberta. http://www.pinchercreekvoice.com/2012/01/grassfires-ravage-areas-of-southern.html (accessed 08.12.13.).

The Vancouver Sun, August 6, 2009. The Mount McLean Wildfire is Raging Nearby, but Claude Terry and his Family are Staying Put. Retrieved from: http://www.canada.com/vancouversun/news/westcoastnews/story.html?id=0596a4e4-fc74-4321-94f1-6cb921f6e3e1 (accessed 26.11.13.).

Turetsky, M.R., Donahue, W.F., Benscotter, B.W., 2011a. Experimental drying intensifies burning and carbon losses in a northern peatland. Nat. Commun. 2, 514. http://dx.doi.org/10.1038/ncomms1523. PMID:22044993.

Turetsky, M.R., Kane, E.S., Harden, J.W., Ottmar, R.D., Manies, K.L., Hoy, E., Kasischke, E.S., 2011b. Recent acceleration of biomass burning and carbon losses in Alaskan forests and peatlands. Nat. Geosci. 4 (1), 27−31. http://dx.doi.org/10.1038/ngeo1027.

Turunen, M.R., Tomppo, E., Tolonen, K., 2002. Estimating carbon accumulation rates of undrained mines in Finland—application to boreal and subarctic regions. Holocene 12, 69—80.

Tymstra, C., Flannigan, M.D., Armitage, O.B., Logan, K., 2007. Impact of climate change on area burned in Alberta's boreal forest. Int. J. Wildland Fire 16, 153—160.

Tymstra, C., 2013. Economic impact of the 2011 fire season in Alberta: a preliminary assessment. In: Presentation to the PRAC II Economics of Adaptation Inter-provincial Forum, March 19—20, 2013. Alberta, Calgary. http://prairiesrac.com/2013presentations/ (accessed 08.12.13.).

Viereck, L.A., 1983. The effects of fire in black spruce ecosystems of Alaska and north Canada. In: Wein, R.W., MacLean, D.A. (Eds.), The Role of Fire in Northern Circumpolar Ecosystems. John Wiley and Sons, New York, pp. 65—80. SCOPE 18.

Wein, R.W., 1976. Frequency and characteristics of arctic tundra fires. Arctic 29 (4), 213—222.

Wells, E.D., Zoltai, S., 1985. The Canadian system of wetland classification and its application to circumboreal wetlands. Quilo. Ser. Botanica 21, 45—52.

Westhaver, A., 2006. Firesmart-forestwise: Managing Wildfire and Wildfire Risk in the Wildland/ Urban Interface. University of Calgary.

White, C.A., Perrakis, D.C.D.B., Kafka, V.G., Ennis, T., 2011. Burning at the edge: Integrating biophysical and eco-cultural fire processes in Canada's parks and protected areas. Fire Ecol. 7 (1), 74—106.

Wotton, B.M., Stocks, B.J., 2006. Fire management in Canada: Vulnerability and risk trends. In: Hirsch, K.G., Fuglem, P. (Eds.), Canadian Wildland Fire Strategy: Background Syntheses, Analyses, and Perspectives, pp. 49—55. Available from: http://cfs.nrcan.gc.ca/pubwarehouse/ pdfs/26529.pdf (accessed 26.11.13.).

Wright, H.A., Bailey, A.W., 1982. Fire Ecology: United States and Southern Canada. John Wiley & Sons, Inc., New York.

Wright, H.E., Heinselman, M.L., 1973. The ecological role of fire in natural conifer forests of western and northern North America: Introduction. Quat. Res. 3 (3), 317—328.

Current Wildfire Risk Status and Forecast in Chile: Progress and Future Challenges

Miguel Castillo Soto, Guillermo Julio-Alvear and Roberto Garfias Salinas

Forest Fire Laboratory, University of Chile, Santiago, Chile

ABSTRACT

The chapter describes the current wildfire risk in Chile and the forecast mechanisms in place for fire management procedures. The country has a high seasonality of fire occurrence, linked to a Mediterranean climate and a highly concentrated spatial distribution of the main areas affected by fire. This knowledge is used to evaluate the meteorological risk at a national level, as well as the possibility of simulating forest fires for prediction purposes, using software developed in Chile and underpinned by a geographic database. The current challenge is to improve the index and to study the phenomenon of severity linked to potential fire impact in greater depth.

4.1 INTRODUCTION

Chile is a South American country with a clear tradition of forestry which is constantly affected by forest fires. Chile covers a territory of approximately 13.6 million hectares of natural woodland and another 2.2 million hectares of commercial forestry plantations. Its Mediterranean-type ecosystem alongside clear seasonality (dry periods in summer) and the human factor provide the necessary conditions for the start and spread of fire and in recent years this has led to considerable damage, including fatalities.

The national territory is characterized by a pronounced spatial concentration of wildfires. Studies by the Forest Fire Laboratory at Chile University's Woodland and Environmental Management Department at the Faculty of Forestry Science and Nature Conservation, combined with records of past fires reported by the National Forestry Corporation (CONAF), have highlighted the existence of particularly critical areas of occurrence and as a

Wildfire Hazards, Risks, and Disasters. http://dx.doi.org/10.1016/B978-0-12-410434-1.00004-X

59

consequence at risk of wildfire. The central and southern areas of Chile are the zones most impacted in terms of the number of fires and surface area affected (Figure 4.1).

A revision of wildfire risk distribution identified a clear wildfire density concentration in the areas close to main population centers (Figure 4.2), such as Valparaíso, Santiago, Temuco, and Concepción. The presence of extensive wildland—urban interface areas is a constant and increasing problem for fire spread risk and danger and has caused a predominance of very aggressive fires and alert levels, which have caused establishment of certain protocols regarding firefighting resource allocation. These fires normally spread rapidly, threatening homes and businesses in the wildland—urban interface areas. Occurrence of risk factors differs from region to region, due to the significant effect of the human population on the initiation of fires, mainly in the peripheral urban areas, along roads and paths where there is an accumulation of combustible vegetation debris and garbage, and also due to the variability of environmental factors that affect the start of fires (climate, topography, and types of vegetation). It is important to point out that even if it is true that practically 99 percent of all fires start as a result of human activity, their characteristics vary from one region to another, including on a very localized geographic scale.

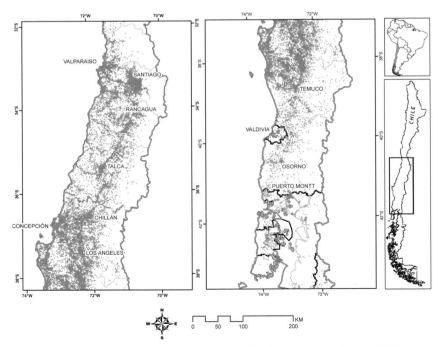

FIGURE 4.1   Spatial distribution of forest fires in Chile. Reference period: 1993—2013. *Forest Fire Laboratory in Chile. Areas marked in red correspond to the numbers of fires coinciding with main population centers and roads.*

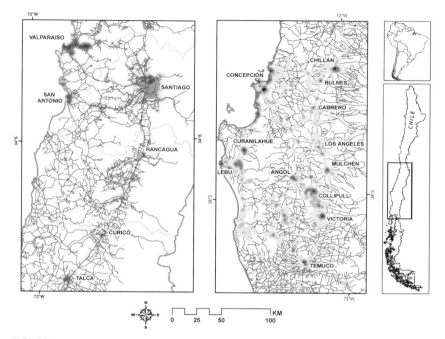

FIGURE 4.2   Location of the main areas at risk of wildfire, defined by critical areas with a past history of fire. Highest densities are shown in darker shades. Information processed by the Forest Fire Laboratory of the University of Chile.

In the last 50 years (1963—2013) 202,601 wildfires occurred in the whole of Chile, which affected 2,270,114 ha. Figure 4.3 illustrates the temporal distribution of these figures, in terms of the number of fires as well as the area affected. The first figure should only be considered as a general reference, given that reliable background information about the number of fires and area affected has only been available since 1986. Nowadays the statistical system figures are more precise, showing that on an average there are 4,135 fires a year. However, within this average very high variability occurs in affected surface area, which indicates that a large number of fires do not necessarily mean a large number of burnt hectares.

In general, the background data reveal a considerable increase in the number of fires up until the period 1982—1986, at which point the curve stabilizes, with values fluctuating between 5,200 and 7,570 annual fires subsequently and up until the present, with a slight upward trend. However, the trend of fire-affected areas in the same period is different, with alternate benign and critical years and fluctuations from 10,921 (for the 2000—2001 season) to 101,691 (1998—1999 season) hectares per year. The recent season 2011—2012 was one of the most critical, with 91,261 burnt hectares. This state

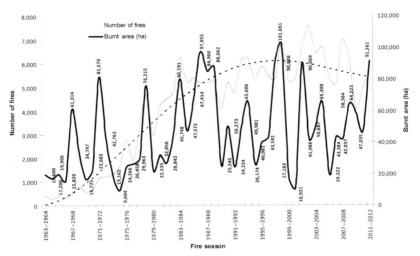

FIGURE 4.3 Historical evolution of fires in Chile over the past 50 years. The dotted line represents the historical trend.

of constant fire occurrence is due to the fact that many forest areas are at high risk of fire and thus fire spread and its subsequent effects.

Wildfire risk evaluation is a topic that has been important in fire management since its origins in Chile more than 40 years ago. This is because decisions about prevention, presuppression, firefighting, and fire use must be based on resource allocation and on evaluation of weather variables that enable wildfires to be anticipated in some way. Therefore research on this topic is underway in Chile, in terms of design and application of indices which permit the wildfire risk to be evaluated and predicted. This text discusses the most relevant issues which explain the historical analysis and the current status for risk and prediction of wildfire.

The effect of prevention campaigns and greater controls regarding fire use should not be overlooked, as these have intensified over the years. Nevertheless, the upward trend of wildfires is a phenomenon that has occurred in most countries, as a consequence of the increasingly intensive use of renewable natural resources, whether for production or as a source of recreation and entertainment. Even though it is not possible to verify this convincingly with scientific evidence, it appears that climate change also influences the increased amount and severity of wildfires.

4.2 INITIAL REFERENCES TO FIRE

The first known initiative on this topic in Chile is the *Wildfire Index* compiled in 1962 by the Forestry Institute (Julio, 1970). It is based exclusively on relative air humidity values, represented in a scale of four ranges: *high*, with a value below 40 percent, *medium* with values between 40 and 60 percent, *low* with values between 60 and 80 percent, and *zero* with values above 80 percent. It was created without the support of any study and in reality its objective is unknown. Its use extended across 2 years and supported prevention campaigns (Julio, 1990). Subsequently in 1968, the University of Chile compiled the *Occurrence Probability Index* (Julio y Sanino, 1968; Ferreira, 1970). This was based on a study of the correlation between various meteorological factors and daily incidences of fire, featuring as a test zone the province of Bío-Bío, located in the central–south area of Chile. Statistical analysis showed that the combination of values of air temperature, relative humidity, and drought factor (based on the number of days without rain) relate reliably to fire probability. In subsequent years, wind velocity was added to this index. Nevertheless, Chile's extreme geographical variability means that the results of this new index are not reliable enough to be used formally with this latest variable.

4.2.1 Relationship between Risk and Danger

Wildfire risk may be considered as a group of indicators that tries to estimate the potential for starting wildfire, fire development, and the consequences of fire spread. At a higher level, risk is inserted into a more detailed mechanism called *degree of danger*, which corresponds to a combination of fixed and variable factors determining the likelihood of fire, fire behavior at the start of the fire, fire spread potential, effort needed for fire control, and the damage that could occur (Julio, 2013).

The quantitative or qualitative explanation of the degree of danger is calculated using the *Degree of danger indices*, which were compiled in accordance with variables and factors most relevant to the start and spread of fire in the region where they will be applied. Given that they evaluate partial aspects of the degree of danger, this approach provides more specific insights into risk in a given area (Julio, 2013). Most of the indices are designed to meet specific objectives that support fire management procedures. It is therefore possible to identify the following types of indices:

Risk index: also known as occurrence or ignition indices, whose purpose is to estimate the probability of wildfire in a defined region and at a specific time.

Drought index: also known as accumulated humidity index, which allows estimation of the vegetation status or its susceptibility to ignition, or its flammability. This summarizes the effect of accumulated drought or precipitation on fire risk.

Propagation index: which makes it easier to estimate the probable speed and range of spread in defined combustibles and in accordance with the environmental conditions prior to the fire.

Intensity index: refers to the estimation of calorific energy emitted by a fire spreading in a particular fuel type and for specific environmental conditions.

Load index: allows estimation of the amount of work or effort required to control a fire.

Total load index: refers to the total amount of daily work needed to control fires occurring in a specific region.

All these indices should function as an interconnected structure whose central objective is the global evaluation of the potential issues caused by wildfires and wildfire spread. Such a structure is known as the *Degree of Danger System*. Given the importance in Chile of such indicators, in 1983 and with the support of the National Research Programme for Fire Management driven jointly by the CONAF, forestry companies, and universities, a program of action was defined with the aim of designing a National Degree of Danger Evaluation System. This comprised four indicators: risk, spread, combustion, and severity (Julio, 1987). Of these four, only the first was completed satisfactorily. Design of these indices concluded in 1989 and was validated in 1990 (Julio, 1990). Final results were expressed in a general formula that sets out the probability of wildfires starting in different sectors of the national territory.

This development in fire prediction has made possible four processes essential to the measurement of fire risk: evaluation of the atmosphere and its possible effect on the occurrence and spread of wildfire, estimation of the current and short-term issues resulting from fire ignition and behavior, calculation of the probable level of conflict of these potential problems, and the possible location of these.

With regard to the application of the risk indices for Chile, the following are highlighted:

Publicity: Sharing information raises awareness of the probable level of wildfire occurrence and spread and encourages the public to collaborate in prevention campaigns.

Control of High-Risk and Danger Zones: Directs the application and monitoring of measures such as suspension of rural work, regulated transit, and forbidden entry to potentially dangerous areas.

Controlled burnings: Local indices allow the efficient regulation of fire use permits in agrosilvopastoral areas and the opportunity to use this measure.

Presuppression: Allows available resources to be programmed, regulated, and reallocated for various procedures, such as detection and combat.

Firefighting strategies: The estimation of the potential behavior of a fire in a defined sector is useful for deciding the most appropriate resources, organization, and combat methods.

Allocation: May represent a fundamental support when deciding the allocation of resources for combat, especially in the case of simultaneous fires.

Planning: Information about past fires from various seasons regarding the degree of danger and showing the chronological and spatial distribution of fire occurrence, as well as enabling strategic planning of fire management both spatially and temporally.

4.3 WILDFIRE RISK INDEX DESIGNED FOR CHILE

In 1990, a global risk index and 15 indices specific to the zones most affected by wildfire for that period were proposed (Julio, 1990). Thus a detailed compilation and analysis of data in the central—southern area of Chile between the months of July 1988 and August 1989 was undertaken, which considered the following stages: (1) definition of the risk areas in the study zone, by consideration of historical information on climate, topography, vegetation, fire occurrence, and population density; (2) detailed collection of past data on wildfire and classification according to risk zone; (3) collection of daily meteorological data in weather stations in each of the risk zones; (4) evaluation of the effect of drought and seasonality and design of their respective roles; (5) definition of variables and evaluation of their weighting for wildfire risk; (6) design of risk indices; and (7) indices validation. The following sources of information were taken into account in the completion of this study:

Climate: Average values of climate variables in Chile over a 30-year period.

Meteorology: Daily meteorological records for four seasons (1985—1988) in 33 weather stations distributed across the whole study area, belonging to Fuerza Aérea de Chile, Armada de Chile, the CONAF, Austral University, the University of Concepción, the Institute for Agro-pastoral Research, and the companies Forestal Arauco, Forestal Celco, and Forestal Mininco S.A.

Wildfire occurrence: Records of wildfire located in the CONAF's fire management statistical system.

Vegetation: Chilean vegetation maps.

Topography: General historical information published by the Military Geographical Institute (the national mapping agency) in Chile's Geographic Atlas.

Population density: Values of the communes of the studied area, provided by the National Institute for Statistics.

The definition of risk zone involved evaluation of variables such as average rainfall for autumn, spring, and summer; average and maximum temperature for January; average annual cloud cover; number of dry months per year; average relative humidity in January; wildfire density; population density; general vegetation distribution; and general topographical distribution. The process involved assigning a standardized score to each of the variables for climate, fire density, and population density in line with the value dispersions of each and their relative weighting for wildfire occurrence probability. Average weighted scores for each variable and each territorial unit were then calculated, allowing a final score to be calculated for each of the territorial units. It was then possible to preliminarily identify the risk zones, using a ranking scale. This definition process took into account historical data about vegetation and topography to identify risk zones, which considered homogeneity in relation to wildfire occurrence and effects of climate, vegetation, and topography. The risk zones identified are presented in Figure 4.4. The design and application of these forecast indices comprise the main reference currently used in Chile. The equations are constantly revised and updated on a regional and national basis, depending on the geographical scale and degree of application. Since 2013 this information is currently being adapted for new geographical information technologies (GIS).

The 15 wildfire risk zones cover a large part of Chile's forestry territory. However, there remain additional large areas in the south for which risk equations still need to be compiled. The risk index calculation results are organized into a general equation and other specific equations for each risk zone. The general risk expression at a national level is as follows:

$$y = 17.6653 + 1.1692x_1 - 0.4378x_2 + 0.3473x_3$$
$$+ 18.6862x_4 - 0.2664x_5 \tag{4.1}$$

where:

$y =$ Probability of daily wildfire occurrence on a scale of 0–100,
$x_1 =$ Air temperature, in degrees Celsius,
$x_2 =$ Relative air humidity, expressed as a percentage,
$x_3 =$ Wind speed, in knots,

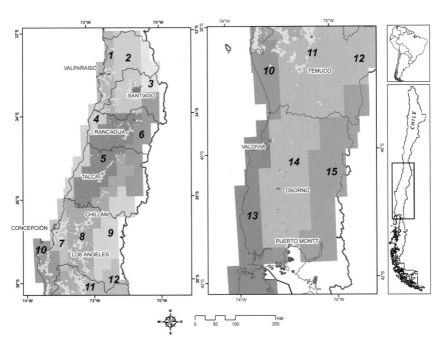

FIGURE 4.4   Wildfire risk zones in Chile. *Forest Fire Laboratory, University of Chile.*

x_4 = Seasonal factor, fluctuating from 1 to 3, and
x_5 = Drought factor, fluctuating from 0.5 to 70.

The application of this index establishes five risk indicators or categories in a specific area: *zero* (0−25), *low* (26−40), *medium* (41−55), *high* (56−70), and *extreme* (71−100). Moreover, local equations per risk area are available, using a formula with the same structure and variables as the general index and differing only in terms of coefficients (Table 4.1). Drought and seasonality factor variables are included in the general index design, as well as the specific indices. Their significance and calculation methods are described here. The average values resulting from this calculation and their ranking in the risk category are indicated in Table 4.2.

4.3.1 Validity of Wildfire Risk Index

It is important to periodically validate the factors involved in the index formation, given the increase in information that must logically occur over time and also bearing in mind that technological developments provide new and improved tools for recording and processing the data required for model formulation and adjustment. On the other hand, wildfire dynamics are a phenomenon with a spatial and temporal variability which must be recorded by all types of indices for accurate representation of fire risk. This is

TABLE 4.1 Risk Index for Chile, Local Equations

Risk Area[a]	Equation
1	$y = 39.9622 - 0.6125X_1 - 0.7252X_2 + 0.7459X_3 + 25.7198X_4$
2	$y = 25.4275 + 0.5078X_1 - 0.7560X_2 - 0.1832X_3 + 37.2736X_4$
4	$y = -9.3043 + 0.9893X_1 - 0.2115X_2 + 2.1037X_3 + 44.0028X_4 - 0.3465X_5$
5	$y = 39.8456 - 1.1176X_1 - 0.7189X_2 + 0.7822X_3 + 67.1255X_4 - 0.2532X_5$
7	$y = 4.833 + 1.2043X_1 - 0.3817X_2 + 0.6190X_3 + 15.6893X_4 + 0.0146X_5$
8	$y = 7.2836 + 0.9017X_1 - 0.3331X_2 + 0.5320X_3 + 27.6157X_4 - 0.2319X_5$
10	$y = 33.9723 + 1.2879X_1 - 0.4123X_2 + 0.2870X_3 + 16.8209X_4 - 0.3631X_5$
11	$y = 15.9073 + 0.9086X_1 - 0.4893X_2 + 0.6895X_3 + 34.5690X_4 - 0.2,848X_5$
13	$y = -63.8288 + 0.5824X_1 - 0.0893X_2 + 1.0241X_3 + 92.1609X_4 - 0.2522X_5$
14	$y = 34.0505 + 1.7960X_1 - 0.4050X_2 + 0.3271X_3 + 8.2711X_4 - 0.3829X_5$

[a]For risk zones 3, 6, 9, 12, and 15, the general equation should be applied.

TABLE 4.2 Risk Categories for Specific Indices

Risk Category (Índex 0–100)

Risk zone	Zero	Low	Medium	High	Extreme
1	0–30	31–50	51–65	66–80	81–100
2	0–25	26–45	46–60	61–75	76–100
4	0–30	31–45	46–60	61–70	71–100
5	0–25	26–40	41–65	66–80	81–100
7	0–25	26–40	41–60	61–75	76–100
8	0–30	31–50	51–65	66–80	81–100
10	0–30	31–45	46–60	61–75	76–100
11	0–30	31–50	51–65	66–80	81–100
13	0–30	31–40	41–55	56–70	71–100
14	0–30	31–45	46–60	61–75	76–100

especially important when human factors are a cause of wildfire. Nevertheless, certain technical aspects must be present for use of this index to continue. The first of these relates to the period of validity. This may be modified by extending analysis for a period beyond the wildfire season. However, this will depend on the wildfire time dynamic recorded over recent years. The second relates to definition of risk areas. At present, a more detailed delimitation of the borders of each polygon represented by the 15 areas is being studied. It relies on an initial database but updates are needed, especially in areas further away from population centers. Nonetheless, the geometric shape of the current risk zones is a satisfactory representation of the problem in central Chile. It is thus highly likely that these initial areas will not change significantly once new data and more advanced SIG-processing techniques are introduced.

4.3.2 Potential Fire Spread Risk and Forecast

The basis for the *Risk* variable ranking in Chile has been explained as being based mainly on meteorological factors and wildfire seasonality. On the other hand, the *Danger variable*, shown in relation to potential fire spread indices, explains the eventual conditions which could result as a consequence of fire spread. Taking account of historical data, in 1993 Chile developed a Wildfire Forecast System based on past scientific records and data from the previous 30 years. This system is called KITRAL (fire, in the Mapuche language) and was developed via a research project funded by the Chile National Scientific and Technological Research Commission, and carried out by a consortium made up of four unities: the Department of Forestry Resource Management at the University of Chile, the Center for Spatial Studies at the University of Chile, the Chile Institute for Technological Research, and the Forestry Institute (Julio et al., 1995).

KITRAL is one of the main technological advances in fire management developed in Chile and is currently used in various applications at national and international levels. In short, it is a Fire Management System whose structure is composed of five large subprograms or modules each with different roles related to fire detection, allocation of firefighting resources, calculation of territorial priorities for fire protection, risk forecast model, and wildfire simulator. In the latter system, design and subsequent construction was followed by successive statistical validation tests (Castillo, 1998) and by continuous improvements, as well as by emerging technological advances in information management and subsequently geographic information systems.

This simulator has been verified in numerous case studies where similarities with KITRAL calculations are corroborated. In general terms, the best models are obtained where the actual area affected is larger because this constitutes a useful prediction tool in terms of potential fire spread and data about past fires is useful to help formulate firefighting strategies, especially for large fires of high calorific intensity. Using the Linear Fire Spread Speed formula proposed in the KITRAL system as a basis, it is possible to determine the time of fire spread from a specific point in the area first burnt and thus the geometric shape of fire expansion. Calculations are made via a geographic information platform and may be applied directly within other computer program settings, such as ArcGIS or Google Earth (Figure 4.5).

One of the most relevant aspects to be considered in fire spread forecast is its application in fires located near the wildland—urban interface. Even when the majority of these are small in terms of affected surface area, the potential for damage is the most serious. As a result, various studies and models have been undertaken into potential fire spread, in light of the presence of population centers within the immediate surroundings of the affected area. When establishing fire protection priority maps, Julio (1992) defined and applied three major components to determine a strategic land planning tool called the *Protection Priority Method*: the first component is *Risk*, associated mainly with the proximity to population centers, paths, agricultural or forestry activities and the number of previous fires. In addition, a second component, *Danger*, refers to the potential conditions for fire spread once it has started. These two important analyses are closely linked to *Potential Damage*, expressed in terms of potential loss of assets and services as a result of fire spread. In summary, the combination of *Risk—Danger—Potential Damage* allows a fire protection priority map to be drawn up, forming a strategic tool for fire protection.

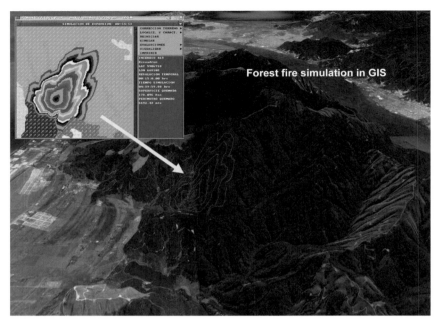

FIGURE 4.5   Simulation of an actual fire in southern Chile. The KITRAL fire spread model in two dimensions is shown in the top left-hand corner. The expanded inset shows the perimeter in three dimensions with Google Earth as a GIS background.

These references are supported by the simulation of those outcomes in which the strategic Danger component is expressed as the potential fire spread resulting from the type of combustible vegetation, slope, and climate variables. These four elements may be recreated using fire spread simulation or forecast and in the case of Chile, this is undertaken using the KITRAL system. Recent references for wildfire simulation in interface areas are listed in Castillo (2013), who developed a study of risk and fire spread areas in the region of Valparaíso in central Chile. Other experiments have been replicated in the outcomes model to plan combat strategies, leading ultimately to a refined economic analysis of firefighting effectiveness. Figure 4.6 shows an advance forecast scheme for an actual fire which affected the wildland belonging to a forestry company and threatened populated areas. Calculation of advance speed, amount of liberated energy, and the geometric characteristics of the spread model (Figure 4.7) provide useful information for developing combat strategies.

4.3.3 Economic Aspects Related to Fire Forecast

Another aspect closely linked to forecast from the point of view of fire management is the economic estimate of actions to be undertaken when faced

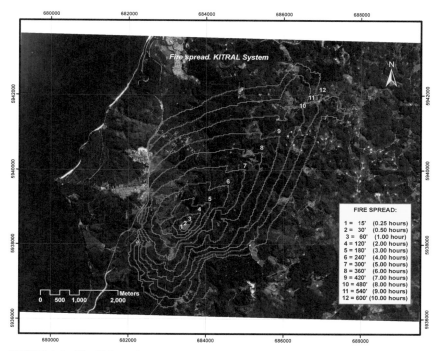

FIGURE 4.6   The 12-hour simulation forecast, for a fire which occurred in southern Chile.

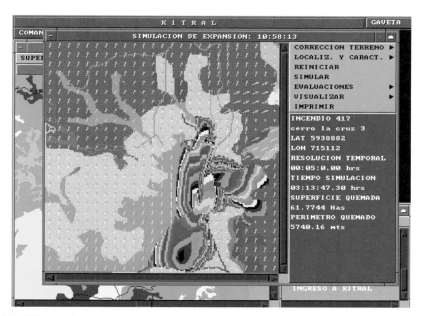

FIGURE 4.7   Simulation of an actual forest fire which occurred in 2008. The forecast for fire spread establishes a reference at each moment for the application of fire suppression strategy.

with fire spread. In Chile, forestry companies and state institutions are constantly concerned with developing better management strategies to deal efficiently with the problem of wildfires.

An example of this is the implementation of protocols for use of land and aerial means to maximize firefighting efficiency while at the same time ensuring they have a reasonably effective result. The former is not easy to achieve, mainly due to the vast number of variables involved in the allocation of methods for fire suppression, based on a forecast which is reliable at the time. Some aspects currently being studied in Chile are described, referring to the integration of variables and criteria for the allocation of firefighting methods based on meteorological and computer simulation forecasts.

For example, one of the latest analyses carried out in the Chile wildfire study looked at suppression costs of a real fire where fire containment was carried out compared to suppression where there was free fire spread. This scenario was studied in terms of economic costs, combinations of the most suitable fire control measures, and the constant support of an updated forecast of fire spread, which is undertaken in Chile using the KITRAL system. This concept is called surface contraction factor (FCS) and corresponds to the quotient between the actual spread when containment procedures are in place and free spread (actual/free) (Table 4.3). This factor provides information about the critical points in the firefighting phases that require greater attention for the decision-making process. This background information is undoubtedly for reference purposes only, because it depends largely on the availability of resources, which are often scarce when active outbreaks happen simultaneously.

TABLE 4.3 Calculation of FCS According to Different Stages of Fire Spread of an Actual Wildfire. Case Study: Wildfire in 2008, Central Chile

Time	Incidence	Burnt Area	KITRAL Forecast	FCS Actual/ Free	Effectiveness (1-FCS)
15:36	Start	0	0	–	–
16:04	First attack	1.7	2.12	0.80	0.20
16:14	Yellow alert	3.18	5.39	0.59	0.41
16:34	Red alert	8.48	10.34	0.82	0.18
16:49	Evacuation maximum severity	13.66	17.77	0.77	0.23
18:51	Burnt houses	20.25	61.77	0.33	0.67

The data in Table 4.3 show the evolution of an actual fire, where the FCS factor was applied. Using the computer simulation, it was possible to thus define the role of costs and losses, depending on the area affected and the duration of the fire spread:

$$M(c + d) = h(a, T) * FCS + d(a, FCS, T) \tag{4.2}$$

In the above expression, M defines the mathematical dependency on the sum of costs of suppression operations (c) and economic losses (d). This function varies in relation to the area affected (a), FCS, and the duration of the fire (T). Bearing in mind that the marginal costs of suppression (h) increase in theory with the area affected, the number of fires (N) directly influences the capacity for land and aerial combat resources. Development of expression (2) using economic simulation results in an average cost which may be identified for each section of land affected at a predicted moment (T) (Castillo, 2013):

$$\frac{\partial M(c + d)}{\partial FCS} = N + \frac{\partial d}{\partial FCS} = 0 \tag{4.3}$$

$$\text{Clarifying } N, \text{ thus gives} : N = -\frac{\partial d}{\partial FCS} \tag{4.4}$$

By incorporating FCS into the analysis when considering the affected area, it is possible to obtain the expression:

$$\frac{\partial M(c + d)}{\partial a} = \frac{\partial h}{\partial a} FCS + \left\{ \frac{\partial d}{\partial FCS} * \frac{\partial FCS}{\partial a} \right\} = 0 \rightarrow = \frac{\partial h}{\partial a} FCS - N \frac{\partial FCS}{\partial a} = 0 \tag{4.5}$$

The point of equilibrium must be found by developing FCS and the marginal costs of suppression (h):

$$\frac{\partial M(c + d)}{\partial (h * FCS)} = 1 + \frac{\partial d}{\partial (h * FCS)} = 0, \quad \text{so that} \frac{\partial d}{\partial (h * FCS)} = -1 \tag{4.6}$$

Developing the relationship of suppression costs allows these calculations to be linked with a combination proposal in the allocation of aerial and land methods of fire control, which establishes both costing by fire severity level and a table of suppression unit costs. These combinations usually fluctuate, especially when concerned with wildfires at the interface. The former may be expressed as a simple mathematic expression which addresses the successive sum of resources j, for a combination of these between aerial and land α, in a time operation threshold t for each size T, and at a suppression unit cost C_u:

$$c_\alpha = \sum_{j=1}^{n} j_i \{ t_T * c_{u_T} * \alpha_T \} \tag{4.7}$$

Although here these relationships have been defined and applied to the context of wildfires in Chile, they are perfectly adaptable for other situations or countries where these types of factors are taken into account in the decision-making process.

4.4 CONCLUSION

The Mediterranean zone of central and southern Chile is characterized by a high occurrence of wildfires, concentrated in particularly critical areas and with a clear seasonality. This is reflected in fire recurrence, fire spread, and damage occurring in Mediterranean countries and areas associated with a communications interface and infrastructure.

As for the meteorological factor and fire occurrence, numerous technical options exist that would improve the wildfire risk index currently used in Chile. However, new research projects in this field obviously need to be promoted. The combined effect of climate change, urban growth, and an increase in fires started deliberately is perceived to increase the severity of fire spread. By revising past records from the past 50 years, it can now be proven that the interface issue is a research priority, since the problem of wildfires differs greatly between regions, even within one location. A great deal remains still to be studied as regards wildfire forecast systems. The ability to simulate future fire spread scenarios as a fire danger prediction tool and the use of historical data to implement more suitable control strategies for different situations is a significant development supporting new geographical information technology.

REFERENCES

Castillo, M., 1998. Método de Validación para el Simulador de Incendios Forestales del Sistema KITRAL. Tesis, U.Chile, Santiago, 122 pp.

Castillo, M., 2013. Integración de variables y criterios territoriales como apoyo a la protección contra incendios forestales. Área piloto: Valparaíso — Chile Central. Tesis Doctoral Ingeniería de Montes. Universidad de Córdoba — España. 247pp.

Ferreira, O., 1970. Estudio de Variables Climáticas como base para la Elaboración de Índices de Peligro de Incendios Forestales en dos zonas de Chile. Memoria, U.Chile, Escuela de Ing. Forestal, Santiago, 58 pp.

Julio, G., 2013. Fundamentos de Manejo del Fuego. Manual Docente. Facultad de Ciencias Forestales y Conservación de la Naturaleza. Universidad de Chile, Santiago, Chile, 300 pp.

Julio, G., Castillo, E., Pedernera, P., 1995. Modelación de Combustibles. In: Actas Taller Internacional sobre Prognosis y Gestión en Incendios Forestales. Proyecto FONDEF FI-13, Santiago, pág: 111—128.

Julio, G., 1992. Método de Determinación de Prioridades de Protección. Docente N° 7. Universidad de Chile, Pub, Santiago, 28 pp.

Julio, G., 1990. Diseño de Indices de Riesgo de incendios forestales para Chile. Revista Bosque 11 (2), 59—72 (Facultad de Ciencias Forestales, Universidad Austral, Valdivia).

Julio, G., 1987. Identificación de líneas y anteproyectos de investigación en manejo del fuego. Inf. Conv. N. Facultad Ciencias Forestales, 28. Universidad Austral, Valdivia.

Julio, G., 1970. Análisis del Control de Incendios Forestales en Chile. Bol.Tecn. N., 32. Escuela Ing.Forestal, U.Chile. Santiago.

Julio, G., Sanino, P., 1968. La Predicción del Grado de Peligro de Incendios Forestales en Chile. Actas IV Jornadas Forestales ACHIF, 218–226. Valparaíso, pág.

Chapter 5

Forest Fires in Europe: Facts and Challenges

Fantina Tedim
Faculty of Arts, University of Porto, Geography Department, Porto, Portugal

Gavriil Xanthopoulos
Hellenic Agricultural Organization "Demeter", Institute of Mediterranean Forest Ecosystems and Forest Products Technology, Athens, Greece

Vittorio Leone
University of Basilicata (retired), Department of Crop Systems, Forestry and Environmental Sciences, Potenza, Italy

ABSTRACT

Forest fires (term used in Europe to designate the unwanted fires burning forests and wild lands) constitute a serious problem for Europe. Frequently, Thought of almost exclusively as a problem for France, Greece, Italy, Portugal, and Spain, this chapter introduces how fire is now a hazard that affects most of the European countries. Although, a sharp gradient exists from the South to the North, in terms of fire regime (e.g., contributing and causing factors, fire frequency and area burned, fire behavior), the problem is common; no country seems exempt. The chapter discusses how contemporary forest fire risk can only be understood from a historical perspective and how this risk is growing exponentially as a result of high and increasing population density and a creeping urban sprawl that is increasing the extent and complexity of the wildland urban interface. This chapter also discusses how significant changes in land-use patterns are conflicting with historical land use practices are affecting the sustainable socioeconomic development in Europe. The impact of changes in critical climatic and weather conditions, such as during recurring heat waves and droughts, is discussed, as is their implications for reconciling social and economic development, environmental concerns, and living with forest fires in a sustainable and dynamic equilibrium in a European context.

5.1 INTRODUCTION

Forest fires (the term used in Europe to designate the unwanted fires burning forests and wild lands) constitute a serious problem throughout Europe. Although there exists a sharp gradient from the South to the North, in terms of fire regime (e.g., contributing and causing factors, fire frequency and area

Wildfire Hazards, Risks, and Disasters. http://dx.doi.org/10.1016/B978-0-12-410434-1.00005-1

burned, fire behavior), no country is exempt. All face growing risk from increasing population density, creeping urban sprawl from incursion into the wildland urban interface (WUI), and from changes in land-use patterns that conflict with societal and ecological protection. Changing climatic and weather conditions are exacerbating these problems. This chapter discusses this complex phenomenon in Europe and highlights the challenges to managing forest fire risk in ways that reconcile social and economic development, environmental concerns, and living with forest fires in a sustainable and dynamic equilibrium.

5.2 FIRE HISTORY: THE EVOLUTION OF FIRE USE IN EUROPE THROUGH PREHISTORY AND HISTORY

Fire is the main driving factor in the story of human evolution (Medler, 2011; Wayman, 2012). Its successive mastering through multiple technologies was a transformative event in human history. In Europe, the use of fire by hominids is reported by James (1989) for 14 sites, dating from 0.4 to 0.6 million years Before Present (Ma BP). Agriculture, or food production as archaeologists call it, appeared in and spread from many different regions of the world between 10,000 and 5,000 years BP, but Neolithic farming spread across Europe from the Middle East between 7,000 and 3,500 years BP (Roberts, 2004), about 4,300 and 2,300 BP for western Europe (Rösch et al., 2004). The Neolithic agricultural revolution used fire to shift the natural vegetation from perennial-dominated to annual dominated landscapes. It has been postulated that people preferred to live in fire-prone places because burning provided them with advantages for hunting, foraging, cultivating, and livestock herding (Pyne, 1995). Fire has been the principal means of forest clearance for agriculture and pastoralism since then (e.g., Rius et al., 2009). Archaeobotanical proxy-data (i.e., pollen and plant macrofossils) give evidence for the early use of fire in agriculture in the northern prealpine lowlands when Late Neolithic (4,300–3,500 BP) settlers used slash and burn (*swidden*; Steensberg, 1957; Conklin, 1961) to convert forest into arable land.

In Nordic countries, where permanent agriculture dates from the late Stone Age (2,500 BP), fire was used to clear land, and to improve grazing quality and wildlife habitats for hunting. Slash and burn played an important role in the economy of northern Scandinavia and large parts of the Finnish inland, during the Late Iron Age and the entire historical period (Alenius et al., 2009) and it was thus the prime means for the colonization of the vast and inhospitable Nordic taiga. Slash and burn practices prevailed in northern countries, in Central Europe, and in Southern countries. It was practiced in Russian provinces until the beginning of the twentieth century, and in Germany (Black Forest) until 1950 (Goldammer, 1998). In Italy, local forms of brutal slash and burning in the forests of *Pinus laricio* were documented until the 1930s in Calabria (Leone and Lovreglio, 2003), a southern region now dramatically characterized by deliberate fire setting. Archival evidence documents that in modern times fire was largely adopted to irreversibly transform the space by pushing back the vegetal cover, mainly forests, thus making room for other forms of land use; we can argue this from specific toponymes. For example, in Italy, the toponyme *cesine* is used to point out a coppice or a forest irreversibly transformed into cultivated land by burning it (Sereni, 1981).

Fire was often documented as an indication of agrarian discontent and a weapon of peasant resistance (Kuhlken, 1999). This has been particularly marked in Sardinia (Italy) (Beccu, 2000) and in Portugal where, during the twentieth century, fire has been used to protest against the afforestation of common lands by a dictatorial government (Devy-Vareta, 1993). Recurrent rural incendiarism is also documented (Harris, 2012) in the island of Cyprus during the period of British governorate and successively Crown Colony (1878–1960). Violent destruction of forests with fire was very frequent and well known in periods of social and political turbulence (Armiero and Palmieri, 2002). Fire history is with a few exceptions (e.g., Amouric, 1992; Araque Jiménez, 1999) a scarcely investigated phenomenon. Concern about forest fires has focused mainly on damages to persons and assets (Jacquot, 1904).

Historically, the presence of vast areas of forest to be colonized, in a time of *hunger for land*, revealed an almost complacent attitude to humans' "victory" over a hostile, powerful nature. For centuries, forest destruction was considered a neutral operation (Vecchio, 1974). People focused on the value of alternative uses for the space the forest occupied: if they burned it, there would be more space for agriculture and farming (Leone and Lovreglio, 2003).

A benefit of adopting a historical perspective is that it affords opportunities to appreciate how fire originated in ways that led to people being able to balance fire use and that have been transmitted through time to become deeply rooted in rural communities in many of the areas now considered fire prone and at high risk. Thus, their nurturing of their traditional ecologic knowledge (Ribet, 2002) has been treated with bans and restrictive regulations rather than integrating them in more farsighted ways to facilitate firewise landscape management.

5.3 FOREST FIRE CURRENT SITUATION

5.3.1 Forest Fire Causes: A Diverse and Complex Reality

In Europe, the causes of forest fires are not yet well understood (Leone et al., 2009; Long et al., 2009; Lovreglio et al., 2006). In 1933, a survey of fires produced by the Institut International d'Agriculture (I.I.A., 1933) showed that fire causes were mainly related to the negligent use of fire in agriculture, railroad sparks, coal kilning in forests, powerlines, and to a much lesser extent, voluntary actions.

The deficiencies in the quality and quantity of information precluded developing accurate and reliable comparisons between countries. Differences among European countries exist concerning (1) the quality of fire-cause investigation, which explains why "uninvestigated" fires remain the prevailing category in forest fire statistics of many countries; (2) the heterogeneity of national classifications, concerning causes of fire categories, the classification criteria, and the level of detail (Camia et al., 2013); (3) the length of time national databases has been available in European countries (most started in the last quarter of the twentieth century); and (4) the too restrictive, former European forest fire classification scheme.

There is a general recognition that knowledge regarding the causes of forest fires is of paramount importance for managing risk, supporting the environmental and civil protection policies, and designing more efficient fire prevention measures (Camia et al., 2013; Catry et al., 2010; FAO, 1999; Ganteaume, 2009; Leone et al., 2009; Long et al., 2009; Montiel and Herrero, 2010). A common Fire Cause classification, to be used by all European countries to report national fire causes to the European Fire Database, was established in 2012 (Camia et al., 2013) and is expected to improve knowledge on this subject. This new hierarchical classification tree identifies 29 fire-cause classes, organized into eight groups and six generic categories (i.e., *unknown*, *natural*, *accident*, *negligence*, *deliberate*, and *rekindle*).

The "unknown cause" class is still the main category in many countries (e.g., France, Germany, Greece, Hungary, Portugal, and Turkey) and is still too frequent in many others (for details see Camia et al., 2013; JRC, 2013). This is explained by the difficulty in determining the ignition point and the cause (Leone and Lovreglio, 2009), or by the high number of fires without information or not investigated (e.g., Portugal and Turkey).

Although statistics on the causes of forest fires in Europe are far from complete, it is evident that the majority of fires are human induced; the Mediterranean region accounts for the larger proportion of human-caused fires in the world (95%) (Leone et al., 2009). "Natural causes" (i.e., caused exclusively by natural processes) represent only a small percentage of all fires and are mostly explained by lightning, which can be very important in some areas of several European countries (e.g., Pausas and Vallejo (1999) for Eastern Spain; Kailidis and Xanthopoulos (1991) for Greece; García-Ortega et al. (2011) for Castile and Leon, Spain).

The anthropogenic causes of forest fires are numerous and complex, and differ between countries and at the regional scale within the same country. With the exception of Spain, Italy, and Poland where deliberate fires (i.e., caused intentionally by people) are the most representative cause, in all the other European countries, forest fire outbreaks result from human activities without intention to provoke damage (i.e., they start by accident, negligent actions, or risk behavior).

Forest fires started by "accident" (i.e., indirectly caused by humans without the use of fire) are mainly related to failure of electric lines, sparks emitted by train brakes or fall of catenaries, sparks emitted by engines and machinery, or firing and explosion during military exercises, etc. In general, they are responsible for the outbreak of a small percentage of fires.

On the contrary, fires caused by "negligence" (i.e., unintentionally caused by people using fire or glowing objects) are more common and are the main cause of fire ignitions in most European countries. Negligent fires are mainly related to agricultural burnings to remove unwanted biomass and to vegetation management for forestry and pasture purposes (e.g., burning of slash or of piles of vegetal waste, and vegetation renovation), because it is a cheap and efficient tool (Leone et al., 2003). The high percentage of fires caused by negligence can be explained by the loss of the traditional knowledge of fire management because of depopulation (Leone et al., 2003; Lovreglio et al., 2010) and the establishment of legal restrictions or prohibiting the use of fire. The increase in the average age of farmers, the aging of rural population, has also amplified the risk associated with traditional fire use practices (Ganteaume et al., 2013; Lovreglio et al., 2010). In some countries (e.g., Germany and Slovakia: JRC, 2013; Lithuania: Camia et al., 2013; Latvia and Netherlands: Ganteaume et al., 2009), fire outbreaks are increasingly caused by recreational activities (e.g., setting barbecues and campfires picnicking, hiking, and hunting) justified by changes in life patterns, by the increasing mobility of people and spread of tourism (Ganteaume et al., 2013). This cause of fire outbreak is also important in Finland, Sweden, and Poland. Fires caused by smoking are of particular importance in Finland, Switzerland, and Italy. Fires originating from accidental or negligent causes usually have some specific features of repetition, concentration, distribution pattern, and relationship with human seasonal activities (Leone et al., 2003). Developing a good understanding of these causes can inform fire-awareness campaigns and consequently reduce the number of ignitions.

Finally, deliberate fires are linked to different motivations, which are very poorly understood. From the available data, it is possible to find significant differences between countries. Setting fire for profit is the most important motivation in many countries (e.g., Italy), mainly related to opening or renovating pasture or for hunting. Interests are also related to setting a fire for monetary (e.g., insurance fraud) or nonmonetary profit (e.g., to set a fire to maintain seasonal employment: Leone et al., 2003; Lovreglio et al., 2010; Vélez, 2000). The irresponsible use of fire assumes a relevant expression in some countries and is explained by psychological troubles and mental illnesses (e.g., France) and by entertainment or children's games (e.g., Sweden, Finland, Poland, and Czech Republic).

The heterogeneity of fire causes highlights the need to consider the social paradigm in forest fire risk management and in developing new forest fire management policies. Fire risk is socially constructed and needs to be resolved by social means (Pyne, 2007).

5.3.2 Wildland–Urban Interface and Forest Fires

One of the main contributions to forest fire risk in Europe derives from the development of extensive WUI. This term describes urban settlements encroaching into, or bordering forested areas. In Europe, especially in the south, an accelerated growth of WUI occurred mainly in the last three decades as the result of significant social and economic changes (Xanthopoulos et al., 2012).

The desire of city inhabitants for better living environments, with less air and noise pollution, fueled development in WUI through the expansion of residential developments in the wild lands around cities. Improved road access and a wider availability of cars facilitated this trend (and an increased number of vacation homes and tourist infrastructures). Sometimes, previously "second home" development areas gradually converted to WUI with housing development for permanent residents (Galiana-Martin et al., 2011).

Contemporarily with the development of WUI, the depopulation of the traditional villages in mountain areas contributed to fire-management problems. Firewood collection in the forestlands around these villages decreased sharply, leading to fuel buildup. Olive groves, vineyards, and orchards, etc, which in the past served as a buffer zone between the wild lands and the settlements, were abandoned or poorly maintained, with abandoned plots being overtaken by natural vegetation (Galiana-Martin et al., 2011). These changes increased the potential for fast-moving fires to reach homes, thus endangering people and properties.

Development of extensive WUI in Mediterranean Europe has contributed to a worsening forest fire risk. People moving into these areas were less knowledgeable about fire and its role in nature. Negligence and associated activities (e.g., construction works, garbage burning, car parking on grasses, open barbeques, and children playing with matches) has increased risk. Further, high land values, when combined with weak land-use planning and/or poor law enforcement, have added deliberate arson to fire causes in these areas.

A further problem in WUI in Mediterranean Europe is the existence of diverse interface typologies (isolated housing, scattered housing, dense clustered housing, and very dense clustered housing) (Lampin-Maillet et al., 2010), often in proximity with each other. This urban sprawl has often been unplanned, lacking order, and in some cases, illegal. Further, in such settlements, their inhabitants tend to understand little about the fire risk associated with where they settled. They expect, irrationally, that public firefighting mechanisms will protect them in any case and so do not undertake measures to make their home safer and defensible. Fire-prevention efforts in such areas have as a rule been inadequate; thus, protecting a particular WUI has become a difficult and risky operation for the firefighters.

The problem of fires in WUI is not limited to Mediterranean Europe. For example, village abandonment by young people has taken place at an alarming rate in Eastern Europe. Agricultural cultivation has stopped over huge tracts of land, and nowadays, forest fires often reach and damage villages.

Greece nicely illustrates the evolution of the WUI forest fire problem. Starting from 1981, when the first WUI fire in Greece spread through the northern high-class suburbs of Athens, fires in Attica gradually became commonplace. For example, Mount Penteli, which forms the NE border of the basin of Athens, burned completely or partially in 1995, 1998, 2000, 2007, and 2009. In 2007, which was the worst fire year recorded for Greece, a series of fires with extreme behavior burned down 270,000 ha of forest and agricultural land. Some of them found vulnerable WUI in their path. More than 110 settlements were affected, resulting in the complete or partial destruction of >3,000 homes and death of 80 people. Deaths and disorderly behavior in WUI areas illustrates the consequences of people's poor level of preparedness, both in regard to readying their homes and how they reacted to approaching fires (Xanthopoulos, 2008a).

Concerning the characteristics of WUI fires in Mediterranean Europe, these can be contrasted quite sharply with those in other parts of the world such as Australia, Canada, and the United States, mainly in regard to the vulnerability of structures to the flames. In Europe, the building materials used for housing construction (e.g., reinforced cement, bricks, stone, and clay tiles) make the structures quite fire resistant. The combination of these fire-resistant homes and associated road and street networks, which in Europe is usually quite dense, modifies fire behavior and the related risk to properties and people (Xanthopoulos et al., 2012). The result is that in Europe complete destruction of a WUI has not occurred so far. Home damage is often only partial. The few people who get injured or die are usually caught in the open, either trying to evacuate or to save their property or livestock.

5.3.3 Large Fires in Europe

Large fires have always been the source of the greatest concern as they contribute most of the yearly burned area. Globally, a fire is characterized as large and draws special attention when it grows to more than a few thousand hectares and causes significant damage to human life, infrastructures, and the environment. In Europe, where the size of continuous forests is generally smaller than those in, for example, Siberia or Canada, and the density of villages and other settlements is quite high, the limit for characterizing a fire as large is relatively low. A size of 100–1,000 ha has been considered as the threshold for characterizing a forest fire as large (Dimitrakopoulos et al., 2011; Ganteaume and Jappiot, 2013). Notwithstanding, the potential for larger conflagrations is growing.

A new term, *megafires*, first coined in the United States in the 2000s to describe fires "that exhibit fire behavior characteristics that exceed all efforts at control, regardless of the type, kind, or number of firefighting assets that are brought to bear" (Brookings Institution, 2005: p.3), has emerged in the European fire literature as, over the last two decades, such fires have become part of the European reality (San-Miguel-Ayanz et al., 2013; Tedim et al., 2013).

Large forest fires that caused serious damages were not unknown in Europe in the remote past, but they were really scarce. For example, in 1888, a large forest fire that started in the suburb of Maroussi near Athens in Greece burned 16,000 ha and stopped only when it reached the sea. The great fire of Landes in 1949 in southwest France burned about 50,000 ha of maritime pine and killed 82 firefighters. In Portugal, the much smaller fire of Serra de Sintra that burned 2,660 ha from September 6 to 12, 1966, remained in memory as it threatened historic places and killed 26 firefighters. Between August 8 and 17, 1975, a large forest fire burned 7,418 ha, mainly of pines, in the Lüneburg Heath in West Germany, and killed six firefighters.

The fires of Landes and of the Lüneburg Heath show that significant vegetation fires are possible, under favorable conditions of weather, fuel, and topography, in any part of Europe. However, it is the Mediterranean region of Europe that is mostly affected by fires. Fires of a few hundred hectares up to a 1−2,000 ha have been quite common in the past, with some larger ones occurring here and there. However, this has been changing in the last 2-3 decades.

While the trend of large fires, defined as those >500 ha, has been shown by San-Miguel-Ayanz and Camia (2010) as being stable in the European Mediterranean region, several fire events have become megafires and caused catastrophic damages and loss of life. Most megafires occurred in years in which fire danger reached very high levels for prolonged periods of time (often overwhelming the capability and endurance of available firefighting resources).

Some notable examples of large fires are: (1) the fires of 1994 in Spain. In that year, 10 fires exceeded 10,000 ha (Vélez, 1995); (2) the fires of Penteli mountain on the outskirts of Athens, Greece, in 1995 and 1999. They burned 6,200 and 7,500 ha, respectively, and damaged hundreds of buildings; (3) the fire of El Solsonès, Catalonia, Spain, in 1998. It burned 27,000 ha and damaged 130 homes (Vélez, 1999); (4) the fires of 2007 in Peloponnese, Evia island, and Attica in Greece that within 4 days (August 24−27) burned about 184,000 ha and killed 64 people (Xanthopoulos, 2007a); (5) the fire of NE Attica, Greece, in 2009. It burned 21,000 ha; the fire affected 15 towns and settlements, destroyed 60 homes, and damaged another 150 (Xanthopoulos, 2012).

Ganteaume and Jappiot (2013) investigated the forest fire records in Southern France and found that <1 percent of the fires recorded during this period were ≥100 ha, but accounted for 78 percent of the burned area. Their results showed that high shrubland and pasture covers, high population and minor secondary road densities, as well as dryness in fall to spring were positively linked to the number of large fires, whereas high forest cover, terrain ruggedness, and wetness in fall to spring were negatively linked to this parameter. On the other hand, the burned area was positively correlated to high wildland vegetation cover, especially shrubland, wetness in fall−winter, dryness in summer for prolonged periods, high unemployment rate, and tourism pressure, while it was negatively related to wetness in summer, high farmland and pasture cover, and high population density. Focusing on large fires in Greece, and setting the burned area threshold at 1,000 ha, Dimitrakopoulos et al. (2011) investigated the forest fires statistics database for the 1990–2003 period. They found that fires exceeding this threshold represented only 0.37 percent of the total number of fires and that they usually spread under moderate to low relative humidity (21–40%) in the presence of strong to moderate winds of northern direction.

Analyzing the large fires in the exceptional fire years of 2003 and 2005 in Spain and Portugal and of 2007 in Greece, San-Miguel-Ayanz et al. (2013) found that megafire events occurred independently of the large expenditures on forest fire suppression and increased preparedness in the countries where they took place. The occurrence of several simultaneous fires with high rate of spread resulted in failures in the initial attack on some of the fires that then became megafires. They were controlled only when the weather conditions improved (and that facilitated effective firefighting). They recommended promoting fire prevention-oriented forest management and increased aware-ness of potential extreme fire events to limit the risk of megafires.

In regard to the future, there is serious concern among scientists that global change, as a result of climate and socioeconomic modifications, will adversely affect forest fire risk, not only in Mediterranean Europe but also in Central and Northern Europe (Giannakopoulos et al., 2011; San-Miguel-Ayanz et al., 2013). Countries such as the United Kingdom, which have been unconcerned with forest fires to date, have seen the signs and have started taking planning measures (Jones, 2013). Norway experienced its largest forest fire in the last 100 years near the municipality of Froland in the south of the country. It burned 2,600 ha, mainly of pine forest, between June 9 and 20, 2008. Sweden was surprised by a large fire in the area of Västmanland in central Sweden on July 31st, 2014 and within the next 12 days grew to a size of 15,000 ha. It can only be hoped that early warnings will be seriously taken into consideration; that emphasis will be given to fire prevention and fire-aware forest manage-ment; that countries will act to mitigate climate change; and that future policies, such as the Common Agricultural Policy of the EU, will consider the side effects in relation to forest fire risk.

5.4 FROM FOREST FIRE SUPPRESSION TOWARD FOREST FIRE RISK MANAGEMENT

5.4.1 The Change of Paradigm

During the twentieth century, the main objective of forest fire policy in Europe was keeping fires as small as possible. This was guided by two assumptions: (1) all fires had negative impacts irrespective of the characteristics of the affected landscapes and ecosystems, and whatever the fire intensity; (2) fire severity was related to fire size (i.e., larger fires mean bigger severity). Guided by the fear of public opinion, the need to protect massive investments in forestry and the requirement to have short-term results, the adoption of a fire suppression-centered policy prevailed (as a very narrow policy objective). Fire became a major threat for the society. The success of this policy precluded seeing that initial "achievements" were worsening the fire problem, generating landscapes with increasing fuel buildup.

The main shortcomings of this policy were the sectorial approach and short-term vision. The former did not allow realizing that the socioeconomic changes in the rural world, the building of an urban society with new standards of living and a new relation with the forest and wild lands were also contributing to fuel load and fire ignition, creating more hazardous fire environments and disseminating fire risk behaviors. At the same time, the negative impacts of total fire exclusion on the regeneration and health of some ecosystems were not considered. The latter explained the lack of any provision for what would happen in the long run, the inexistence of reference to the cost, and the unawareness of environmental constraints (Xanthopoulos, 2007b), as well as the dynamic characteristics of fire environment emphasized by local (e.g., changes in forestry and agricultural production systems, WUI expansion) and global changes (e.g., climate change).

It is a common place saying that this fire suppression policy failed, but taking into consideration its first objective, it did not fail because most of the fires recorded in European databases are small (e.g., 71% of the fire events that occurred in Portugal since 1980 burned <1 ha and only 0.9% burned ≥100 ha). It failed in considering itself the panacea to address the forest fire problem in Europe. The obvious response to an increasing number of ignitions and to larger and more intense fires dangerously threatening villages and population was to reinforce suppression capacity through costly equipment, a policy sometimes influenced by industrial/professional lobbies. However, firefighting capacity improvement did not prevent major fire disasters in WUI in Greece (Xanthopoulos, 2007a), or in Portugal (Tedim et al., 2013). It was assumed that forest fires were an emergency and consequently a civil protection or civil defense problem. However, by 1999, the Committee on Forestry from Food and Agriculture Organization of the United Nations (FAO) had already recommended that the solution to the forest fire problem could not be based on a reaction to a situation that has already developed, but rather on the proactive prevention before the emergency arises.

Even though the current European forest fire policy remains imbalanced, with too many resources allocated to suppression compared to investment in prevention (Rego et al., 2011), progress has been made since the end of the twentieth century. In the last decade, Southern European countries established national frameworks to defend forests from fire (e.g., the National Plan for Defense of the Forest against Fires that was published in Portugal in 2006) including significant improvements in prevention, detection, recovery, and law enforcement, in addition to suppression. However, if the presence of these national frameworks clearly identified fundamental fields of action to address forest fires, it does not mean that they have been fully implemented or that they have been used to full advantage. The countries where successful changes are occurring (e.g., France, Spain) have developed holistic approaches, covering prevention and suppression, with a central planning and decentralized operation management (Oliveira, 2005). One of the main lines of supporting evidence of a shift from a reactive policy focused on fire suppression, toward a more proactive policy on fire management, is fire use practice.

5.4.2 Fire as a Management Tool

Currently, there is a diversity of fire use practices in Europe, according to legal framework and management objectives. Fire use is totally or partially forbidden in several countries (e.g., Belgium, Bulgaria, Czech Republic, Germany, Hungary, Poland, Romania, and Slovenia) (Montiel-Molina, 2011). Conversely, fire use is regulated in Southern European countries, Baltic countries, Switzerland, and United Kingdom. In Portugal, and in some regions of Spain, France, and Italy, there is specific regulation on prescribed and suppression fire. The use of fire is not regulated in Denmark, Finland, Norway, and Sweden.

In Northern European countries (e.g., United Kingdom, Sweden, and Finland), fire is mainly used for nature conservation of protected areas and forest regeneration, and at present, its use is required by forest certification schemes (Goldammer et al., 2007; Granström, 2001). Conversely, in Southern European countries, the main purposes of fire use are related to land management: improvement of pasture for grazing, sylviculture improvements, habitat restoration, and management (namely, for hunting purposes), fire-hazard reduction to decrease future fire intensity and severity, and firefighter training.

The traditional fire use (TFU) refers to communities using fire for land and resource management purposes based on traditional know-how (Lázaro, 2010). In the Mediterranean Basin as well as in other European countries of recent integration in the EU (i.e., Bulgaria, Lithuania), the TFU is still alive and is a deeply rooted tool for agricultural and forest stubble and grazing improvement (Montiel-Molina, 2011) although banned by law in many countries. Illegal burning is often carried out as a *hit-and-run* practice (e.g., by pastoralists setting fires and fleeing the site to avoid prosecution) (Goldammer and Montiel, 2010). This restrictive legal framework has contributed to turning TFU into the most important cause of forest fires in many European countries (e.g., Portugal).

In several parts of Europe, fire culture was lost not only because of the rapid socioeconomic changes after World War II but also because of the expansion of the generally prevailing opinion that fire would damage ecosystem stability and biodiversity, which led to fire bans in most European countries (Goldammer, 1998). If there is a general abandonment of traditional use of fire in Central European and Baltic countries, contrarily, its reestablishment is a reality in some regions of Norway, Germany, and Finland (Goldammer et al., 2007; Haaland, 2003; Lovén and Äänismaa, 2004; Montiel-Molina, 2011) as a traditional land management technique for the maintenance of many landscapes and ecosystems, as well as a tool for reducing fuel hazard (Lázaro and Montiel, 2010).

Currently, there are several reasons and approaches to maintain, restore, or introduce the use of fire in land and ecosystem management. Even though since the 1970s some researchers showed the ecological importance of fire in several ecosystems (e.g., Susmel, 1977; Vélez, 1982), the social and the political acceptance of fire as a tool for landscape management has been met by strong resistance. The reestablishment of the use of fire generally converged in the implementation of prescribed burning (PB) which is, with few exceptions, increasingly restricted to trained and specialized personnel of state organizations (Castellnou et al., 2010) or of private enterprises.

PB represents the controlled and careful application of fire to vegetation under specific environmental conditions (i.e., specified fuel and weather conditions) which allow the fire to be confined to a predetermined area and at the same time to produce the intensity of heat and rate of spread required to accomplished predetermined and well-defined management objectives and long-term management goals, established in the planning phase (Castellnou et al., 2010; FAO, 2005; Fernandes et al., 2002). The use of PB to prevent forest fires was already reported in the nineteenth century in French (Delabraze and Valette, 1982) and Portuguese pinelands (Botelho and Fernandes, 1998); however, the first experiences in the use of PB as a professional technique for vegetation management and forest fire prevention go back to the 1970s in Southern European countries (Faerber, 2009; Fernandes et al., 2002; Leone et al., 1999; Liacos, 1977; Vélez, 1982). The main constraints for PB introduction were legislation (e.g., lack of certification programs, fire bans), complex land structure (e.g., WUI, fragmentation of private property), cultural (e.g., fear of fire escapes, lack of professional experience and knowledge of ecological effects of fire, and lack of knowledge in fire behavior forecast), and management policies (e.g., priority to fire control policies) (Ascoli and Bovio, 2013; Leone et al., 1999; Xanthopoulos et al., 2006).

However, under the influence of very large and severe fires, European societies have been slowly but increasingly reconsidering the need to use fire in the form of PB and as a suppression tool (Castellnou et al., 2010). The PB regulation is very different across Europe. In France, Spain, Italy, and Portugal, PB has been incorporated as a management practice, and it is supported by specific legal and legislative tools. In most other European countries, constraints, obstacles, and negative attitudes toward the use of fire remain important factors resulting in restrictive legal frameworks, complex territorial structures, lack of experience among professionals, or negative perception from society (Xanthopoulos et al., 2006).

On the other hand, as more research and experience have been accumulated, the use of PB has been broadened from forest fuel reduction toward more environmentally oriented purposes (i.e., biodiversity and nature conservation) in Southern European countries. In practice, the development of PB programs is carried out with greater or lesser integration of traditional know-how—from a complete implementation by professionals, to implementation by professionals with the support of the communities (e.g., shepherds, farmers), to training of and then implementation of PB by the communities (Lázaro and Montiel, 2010), in any case, capitalizing their traditional ecological knowledge (Ribet, 2002).

The evolution of PB into a collaboration practice between professional teams and local communities (i.e., implementation of PB by professionals with the support of local communities or implementation by local communities after training in the technical aspects of fire behavior) permits enlarging the number of people able to use fire in a sustainable way. It also contributes to smoothing conflicts, because such use of fire is no longer clandestine and people can be proud to exhibit their capacities in using fire in sustainable ways. Such activities are being carried out, with positive results, in Sardinia (Italy) (Delogu, 2013).

Suppression fire (i.e., backfire, back burn, and burnout) (Miralles et al., 2010) is a practice that has been recuperated in several countries as it is a powerful and very efficient approach, offering a wide range of tactical options with a rather long tradition in some European countries, although its use is currently limited, and often clandestine (Montiel et al., 2010). The regulated use of suppression fire is mainly concentrated in Southern Europe (i.e., Cyprus, Portugal, and in some regions of Spain, France, and Italy).

5.4.3 Fire Management Organization

The similarities in terms of the Southern European countries concerning the frameworks used to defend forests from fire are often inconsistent with organizational structures, with political-administrative decentralization and organizational cultures in each country, adding to the complexities of understanding fire risk. In Italy, the State Forestry Corps is a police force responsible for fire prevention and suppression. It has competences in sensitization campaigns, fire-cause investigations, burned-area mapping, fire-occurrence reporting, and fire statistics. In addition, it coordinates the use of water bombers belonging to the State fleet. Suppression of fire, on the contrary, is a responsibility of the Regions, which can get the help of the Forestry Corps through specific agreements, with the exception of some Regions with a special Statute, which have regional corps of forest rangers.

In Portugal, where forestland belongs predominantly to private owners, fire management is shared by several state agencies and is organized at national, district, and municipal levels. Since the establishment of the System of Protection of Forest Against Fire the municipalities have a bigger responsibility in the planning and management of fire risk. Fire management was a competence of the Forest Services until 1980. Forest Services (now called Institute for Nature Conservation and Forests) retained responsibility for the coordination of measures of structural prevention, fuel management, and sensitization campaigns. However, coordination of suppression is a responsibility of the National Civil Protection Authority. The suppression activities are carried out by fire brigades, which most of them are associations of civil society where the highest percentage of work is conducted by volunteers. Fire brigades can also develop detection activities coordinated by the Municipal Service of Civil Protection. The other interveners in fire management are the National Guard (i.e., a military force responsible for the coordination of detection system and law enforcement, which also has initial attack intervention professional teams); forest rangers team (fuel management, PB, detection, and initial attack), which can be established by public or private organizations. Private fire brigades have also been established namely by paper companies. The aerial resources used in fire combat belong to private companies.

In France, Civil Protection is organized in the national level (ministries), the "zonal" level (seven defense zones), and the local level (95 "local circumscriptions" each with a fire and rescue department composed of 30–60 fire brigades being responsible for prevention and fighting namely forest fires). Forest firefighting is the responsibility of the Fire and Rescue Service since the late 1980s and is based on fire brigades. The structure of the organization is military-like. The local fire and rescue departments are financially run by local authorities, but they follow national rules and methods. The fire chief is appointed by both the local circumscription Chairman and the Ministry of Interior (Peuch, 2005).

In Spain, forest firefighting is organized in a different way in each of the autonomous regions and the fire prevention and suppression responsibility scheme varies considerably from one to another. For example, in the highly flammable Andalucia region in the south, the Forest Authorities are responsible for fire suppression. In the less flammable but more forest rich Catalonia, forest fire suppression responsibility lies with the Fire Service. The province of Madrid, where urban areas are numerous and the forests are not as flammable as in Andalucia, follows a mixed model, where the forestry authorities are not completely disassociated from fire suppression (Xanthopoulos, 2008b).

In Greece, approximately 65 percent of the forests and forestlands belong to the state while a significant part of the remaining 35 percent belongs to municipalities and monasteries. As a result, forest fire management has always been a responsibility of the state. Until 1998, the Greek Forest Service was responsible for fire prevention and suppression. Then, the responsibility for fire suppression was transferred to the Greek Fire Brigades. Currently, the Forest Service maintains a role in fire prevention (e.g., prevention planning, forest-fuel management, forest road network maintenance) whereas the General Secretariat for Civil Protection has a coordinating role, including facilitating the supporting role of volunteer groups and of the resources of the local authorities to the suppression activities. The Air Force operates a fleet of state-owned water bombing airplanes. Both this fleet and the 12–15 contracted privately owned helicopters are dispatched by the Fire Brigades coordination center. The effectiveness of this fire management scheme, as it has performed since 1998, is in doubt (Xanthopoulos, 2012).

5.5 THE ROLE OF EUROPEAN UNION POLICIES IN FOREST FIRE MANAGEMENT

The *Council Decision 89/367/EEC*, followed in December 1996 by the Thomas Report on the European Union's forestry strategy, represents a first step in the development of an EU Forest Policy, since no EU Treaty provides for a comprehensive common policy. Soon after that, in the early 1990s, the EU recognized forest fires not only as a problem of the Mediterranean countries but also as a common issue in the rest of the regions.

Protection against forest fires, as a Community action in support of forest conservation and management, has been influenced by a series of successive binding measures (regulations, financial support for actions), which gradually shifted their perspective from concerns for the protection of forests and safety of people and property, particularly in Southern Europe, to the broader scenario of environment and climate change.

In the last decade, forest policy experienced increased attention by EU institutions (the "European Parliament resolution on Natural Disasters" (2005), the EU "Forest Action Plan" (2006), the "Green Paper on forest protection and information" (2010), the "Arsenis Report" (2011)). In the wake of large-scale flooding in Europe, in 2002, and devastating forest fires in the summer of 2005, the European Parliament adopted the "European Parliament resolution on Natural Disasters (fires, droughts, and floods)—environmental aspects (2005/2192(INI))". Enhancement of the protection of EU forests is among the key actions of the *EU Forest Action Plan*, which invites the Member States to support forest fire prevention measures, restoration of forests damaged by natural disasters and fire, studies on the causes of forest fires, awareness raising campaigns, training and demonstration projects. The purpose of the "Green Paper" was to debate on options ensuring that forests continue to perform all their productive, socioeconomic, and environmental functions in the future. The "Arsenis Report" on the Commission Green Paper underlines the importance of forest fire prevention through landscape planning and connectivity, infrastructure and training.

Strictly concerning forest fires, the EU activity developed through measures and relative funding, which gradually focused on fires more as a crucial environmental issue than as a sectorial activity in agriculture, thus passing competence from Agriculture (DG AGRI—Directorate General Agriculture and Rural Development—between 1992 and 2002) to Environment (DG ENV—Directorate General Environment), from 2003 onward.

The "Council Regulation (EEC) No 2158/92 on protection of the Community's forests against fire" (and the regulations No 1170/93, No 804/94, No 1727/1999, No 805/2002), focused on prevention significantly improving knowledge about forest fires, and fostered forest protection by reducing fire outbreaks and burned surfaces. This regulation was replaced by "Regulation (EC) No 2152/2003 of the European Parliament and of the Council concerning monitoring of forests and environmental interactions in the Community (Forest Focus)" with the complementary regulations No 2121/2004 and No 1737/2006.

"Forest Focus" was a Community scheme, running from 2003 to 2007, mainly focused on protecting forests against atmospheric pollution and fires. It covered different areas including prevention of fires and their causes, and supported the implementation of the European Forest Fire Information System (EFFIS). EFFIS services (San-Miguel-Ayanz et al., 2012) include fire danger rating by Fire Weather Index; active fire detection; maintenance update and analysis of a European forest fire database; burned-area mapping; land-cover damage assessment; appraisal of Natura 2000 sites affected; forest fire emissions; potential soil erosion; and estimates of economic losses caused by forest fires in Europe. EFFIS is complementary to national and regional systems, and provides information required for international collaboration on forest fire prevention and fighting and in cases of crossborder fire events (Corti et al., 2012).

5.6 CONCLUSION

Based on the evidence presented here, it is clear that fire has been present in Europe since prehistory. It has played very important roles in the lives of people and was not considered a great menace. As a result of social and economic changes, fire became the main threat to European forests in the second half of the twentieth century. The result was the development of a complete fire-exclusion policy based on very strong, sometimes military-like, fire-suppression organizations. This policy of fire exclusion revealed two main handicaps. First, it has a reactive approach and does not address the roots of the problem. So far, it has promoted the shift of competencies on fire management at the organizational level from Forest Services to Fire Brigades in many European countries (e.g., Greece, Portugal); second, the total fire exclusion increased the fuel load in forests and contributed to the appearance of fires with a very high severity that grow to previously unheard of sizes, regardless of the firefighting resources.

The need for changing this paradigm, giving more importance to fire-aware forest management, and to fire prevention, has been facing institutional inertia and resistance. The increased acceptance of the full range of fire effects (i.e., from very detrimental to very beneficial) as well as the appearance of regulation of fire use in some countries appears as a sign of hope. The very diverse and complex anthropogenic causes of fire point out the need to consider the social and cultural dimensions in any policy that addresses the problem of forest fires.

It is fundamental to move on to new policies that (1) attend to the complex and dynamic interrelations between social, economic, environmental, and political drivers; (2) integrate all public policies that address forest fire related to issues; (3) adopt more collaborative patterns in the relations between institutions and between institution and communities, actively facilitating agreements between stakeholders in territory for common action; (4) develop a centralized, general planning framework aimed at developing balance between prevention and suppression but allowing decentralized operations adapted to the local socioeconomic and environmental characteristics; (5) develop awareness campaigns to prepare societies for a changing paradigm to reduce pressure to the politicians for stronger fire suppression and more use of aerial resources; (6) look to communities as allies and wardens of the territory rather than as adversaries and the eternal culprits; (7) promote rational land use for people and the environment based on sound scientific criteria; and (8) change fire-control competencies from an institutional monopoly of some actors to a far-sighting action of involvement of the whole society occupying a territory.

REFERENCES

Alenius, T., Lavento, M., Saarnisto, M., 2009. Pollen-Analytical Results from Lake Katajajärvi—Aspects of the History of Settlement in the Finnish Inland Regions, 12. Acta, Borealia, 26, 136–155.

Amouric, H., 1992. Le feu à l'épreuve du temps. Editions Narration, Collection Témoins et arguments, 256 p.

Araque Jiménez, E., 1999. Incendios Históricos: Una Aproximación Multidisciplinar. Universidad Internacional de Andalucía, 422 p.

Armiero, M., Palmieri, W., 2002. Boschi e rivoluzioni nel Mezzogiorno. La gestione, gli usi e le strategie di tutela nelle congiunture di crisi di regime (1799–1860). In: Lazzarini, A. (Ed.), Diboscamento montano e politiche territoriali. Alpi e Appennini dal Settecento al Duemila. Franco Angeli Storia, Milano, 598 P.

Ascoli, D., Bovio, G., 2013. Prescribed burning in Italy: issues, advances and challenges. Iforest-Biosciences and Forestry 6, 79–89 [online 2013-02-07] URL: http://www.sisef.it/iforest/contents?id=ifor0803-005.

Beccu, E., 2000. Tra cronaca e storia le vicende del patrimonio boschivo della Sardegna. Carlo Delfino editore, Sassari, 432 p.

Botelho, H.S., Fernandes, P.M., 1998. Controlled burning in the mediterranean countries of Europe. In: Proceedings Advanced Study Course on Forest Fire Management, 6–10 October 1997. Marathon, Athens, Greece (Algosystems, DGXII/CEE).

Brookings Institution, 2005. The Mega-Fire Phenomenon: Toward a More Effective Management Model. Concept Paper. Presented to the U.S. National Fire and Aviation Board, 20th September, 2005. The Brookings Institution Centre for Public Policy Education, Washington, D.C, 15 p.

Camia, A., Durrant, T., San-Miguel-Ayanz, J., 2013. In: Harmonized Classification Scheme of Fire Causes in the EU Adopted for the European Fire Database of EFFIS Executive Report. JRC, European Commission, Ispra, Italy.

Castellnou, M., Kraus, D., Miralles, M., 2010. Prescribed burning and suppression fire techniques: from fuel to landscape management. In: Montiel, C., Kraus, D. (Eds.), Best Practices of Use-Prescribed Burning and Suppression Selected Case-study Regions in Europe. European Forest Institute Research Report 24, Joensuu, pp. 3—16.

Catry, F.X., Rego, F.C., Silva, J.S., Moreira, F., Camia, A., Ricotta, C., Conedera, M., 2010. Fire starts and human activities. In: Silva, J.S., Rego, F., Fernandes, P., Rigolot, E. (Eds.), Towards Integrated Fire Management—Outcomes of the European Project Fire Paradox. European Forest Institute Research Report 23, Joensuu, Finland, pp. 9—22.

Conklin, H.C., 1961. The Study of shifting cultivation. Curr. Anthropol. Vol. 2 (1), 27—61.

Corti, P., San-Miguel-Ayanz, J., Camia, A., McInerney, D., Boca, R., Di Leo, M., 2012. Fire news management in the context of the European forest fire information system (EFFIS). In: Quinta conferenza italiana su lsoftware geografico e sui dati geografici liberi (GFOSS DAY 2012), November 2012.

Delabraze, P., Valette, J.C., 1982. The use of fire in silviculture. Gen. Tech. Rep. PSW-58. In: Conrad, E., Oechel, cords, W.C. (Eds.), Proceedings of the Symposium "Dynamics and Management of Mediterranean-Type Ecosystems. Pacific Southwest Forest and Range Experiment Station, Forest Service, U.S. Department of Agriculture, Berkeley, CA. pp. 475—463.

Delogu, M.G., 2013. Dalla parte del fuoco. Ovvero il paradosso di Bambi. Edizioni Il Maestrale, 207 p.

Devy-Vareta, N., 1993. A floresta no espaço e no tempo em Portugal. A arborização da Serra da Cabreira (1919—1975). Universidade do Porto, Porto, 459 p.

Dimitrakopoulos, A.P., Vlahou, M., Anagnostopoulou, ChG., Mitsopoulos, I.D., 2011. Impact of drought on wildland fires in Greece; implications of climatic change? Clim. Change 109, 331—347.

Faerber, J., 2009. Prescribed range burning in the Pyrenees: from a traditional practice to a modern management tool. Int. For. Fire News 38, 12—22.

FAO, 1999. In: Report of Meeting on Public Policies Affecting Forest Fires. Fourteenth Session, 1999 March 1—5; Rome, Italy.

FAO, 2005. Wildland Fire Management Terminology. FAO. Forestry Paper, No. 70, 257 p.

Fernandes, P., Botelho, H., Loureiro, C., 2002. Manual de formação para a Técnica do Fogo controlado. Departamento florestal, Universidade de Trás-os-Montes e Alto Douro, Vila Real.

Galiana-Martin, L., Herrero, G., Solana, J., 2011. A wildland—urban interface typology for forest fire risk management in Mediterranean areas. Landscape Res. 36 (2), 151—171.

Ganteaume, A., 2009. Détermination des causes d'incendie de forêt et harmonisation desméthodes pour les rapporter. Info DFCI, Bulletin du Centre de Documentation ForêtMéditerranéenne et Incendie, CEMAGREF 63, 6.

Ganteaume, A., Camia, A., Jappiot, M., San-Miguel-Ayanz, J., Long, M., Lampin, C., 2013. A review of the main driving factors of forest fire ignition over Europe. Environ. Manage. 51 (3), 651—662.

Ganteaume, A., Jappiot, M., 2013. What causes large fires in Southern France. For. Ecol. Manage. 294, 76—85.

Ganteaume, A., Jappiot, M., Long, M., Lampin-Maillet, C., Duché, Y., Savazzi, R., Bonora, L., Conese, C., Piwnicki, J., Ubysz, B., Szczygiel, R., Galante, M., Ferreira, A., Suarez-Beltran, J., 2009. State of the Art (Draft).Deliverable D 1.1.Contract Number 384 340 "Determination of Forest Fire Causes and Harmonization for Reporting Them. European Commission-JRC, 220 p.

García-Ortega, E., Trobajo, M.T., López, L., Sánchez, J.L., 2011. Synoptic patterns associated with wildfires caused by lightning in Castile and Leon, Spain. Nat. Hazards Earth Syst. Sci. 11, 851—863.

Giannakopoulos, C.A., Hatzaki, M., Karali, A., Roussos, A., Xanthopoulos, G., Kaoukis, K., 2011. Evaluating present and future fire risk in Greece. In: San-Miguel-Ayanz, J., Gitas, I., Camia, A., Oliveira, S. (Eds.), Proceedings of the 8th International EARSeL FF-SIG Workshop on "Advances in Remote Sensing and GIS applications in Forest Fire Management — From local to global assessments" 20—21 October 2011, Joint Research Centre of the European Commission, vol. 24941. Institute for Environment and Sustainability, EUR, Stresa (Italy), pp. 181—185.

Goldammer, J.G., 1998. History of forest fires in land-use systems in the baltic region: implications on the use of prescribed fire in forestry, nature conservation and landscape management. In: Proceedings, First Baltic Conference on Forest Fires, Radom-Katowice/Poland 1998 (in Polnisch mit englischen Abstracts), Narodowy Fundusz Ochrony Srodowiska i Gospodarki Wodnej. Forest Research Institute, Warsaw, pp. 59—76.

Goldammer, J.G., Hoffman, G., Bruce, M., Kondrashov, L., Verkhovets, S., Kisilyakhov, Y.K., Rydkvist, T., Page, H., Brunn, E., Lovén, L., Eerikäinen, K., Nikolov, N., Chuluunbaatar, T.O., 2007. The Eurasian fire in nature conservation network (EFNCN): advances in the use of prescribed fire in nature conservation, landscape management in temperate-boreal Europe and adjoining countries in Southeast Europe, Caucasus, Central Asia and Northeast Asia. In: Proceedings of the 4th Internacional Wildland Fire Conference, pp. 13—17. Seville, Spain.

Goldammer, J.G., Montiel, C., 2010. Identifying good practices and programme. Examples for prescribed burning and suppression fire. In: Montiel, C., Kraus, D. (Eds.), Best Practices of 3*' Use-Prescribed Burning and Suppression Programmes in Selected Case-Study Regions in Europe. European Forest Institute, Joensuu, pp. 35—44. Research Report 24.

Granström, A., 2001. Fire management for biodiversity in the European boreal forests. Scand. J. For. Res. 3, 62—69.

Haaland, P., 2003. Five thousand years of burning. The Eur. Heathlands, 165.

Harris, S.E., 2012. Cyprus as a degraded landscape or resilient environment in the wake of colonial intrusion. Proc. Nat. Acad. Sci. 109 (10), 3670—3675.

I.I.A. (Institute International d'Agriculture), 1933. Enquête Internationale sur les incendies de forêts. Institute International d'Agriculture, Rome, 457 p.

Jacquot, A., 1904. Incendies en forêt. Evaluation des dommages. Nancy, Berger-Leyrault & C, Paris ie, Editeurs, 412 p.

James, S.R., 1989. Hominid use of fire in the lower and middle Pleistocene: a review of the evidence. Curr. Anthropol.(Univ. Chicago Press) 30 (1), 1—26.

Jones, M.S., 2013. Merging land management with emergency management—practical steps and challenges. Seoul Education Cultural Center, Seoul, Republic of Korea. In: Proceedings of the International Symposium on Strategy Development of Forest Fire Policy and Organization. January 15—17, 2013. Korea Forest Research Institute, Seoul, pp. 27—37.

JRC, 2013. Forest Fires in Europe Middle East and North Africa 2012. Publications Office of the European Union, Luxembourg, 109 p.

Kailidis, D., Xanthopoulos, G., 1991. The Forest Fire Problem in Greece. Aristotelian University of Thessaloniki. Department of Forestry and Natural Environment, Forest Protection Laboratory, Greece,. No. 3.10 p.

Kuhlken, R., 1999. Settin' the woods on fire: rural incendiarism as protest. Geogr. Rev. 89 (3), 343—363.

Lampin-Maillet, C., Jappiot, M., Long, M., Bouillon, C., Morge, D., Ferrier, J.P., 2010. Mapping wildland—urban interfaces at large scales integrating housing density and vegetation aggregation for fire prevention in the South of France. J. Environ. Manage. 91, 732—741.

Lázaro, A., 2010. Development of prescribed burning and suppression fire in Europe. Research Report 24. In: Montiel, C., Kraus, D. (Eds.), Best Practices of Use-Prescribed Burning and Suppression Selected Case-Study Regions in Europe. European Forest Institute, Joensuu, pp. 17—31.

Lázaro, A., Montiel, C., 2010. An overview of policies and practices related to fire ignitions at the European Union level. In: Sande, J., Rego, F., Fernandes, P., Rigolot, E. (Eds.), Towards Integrated Fire Management—Outcomes of the European Project Fire Paradox. European Forest Institute, Finland, pp. 35—46. Research Report 23.

Leone, V., Koutsias, N., Martínez, J., Vega-García, C., Allgöwer, B., Lovreglio, R., 2003. The human factor in fire danger assessment. In: Chuvieco, E. (Ed.), Wildland Fire Danger Estimation and Mapping. The Role of Remote Sensing Data. World Scientific Publishing, Singapore, pp. 143—196.

Leone, V., Signorile, A., Gouma, V., Pangas, N., Chronopoulous-Sereli, A., 1999. Obstacles in prescribed fire use in Mediterranean countries: early remarks and results of the fire torch project. In: Proceedings of the "DELFI International Symposium. Forest Fires: Needs and Innovations" Athens (Greece) 18—19 November 1999, pp. 132—136.

Leone, V., Lovreglio, R., 2003. Human fire causes: a challenge for modeling. In: Chuvieco, Emilio, Martín, Pilar, Justice, Chris (Eds.), Innovative Concepts and Methods in Fire Danger Estimation. Proceed. 4th International Workshop on Remote Sensing and GIS Applications to Forest Fire Management: Ghent University Ghent—Belgium, 5—7 June 2003, pp. 91—99.

Leone, V., Lovreglio, R., 2009. Il fuoco come strumento nella gestione preventiva e tattica (Fondazione S. Giovanni Gualberto—Osservatorio Foreste e Ambiente ed). I Quaderni 9. In: La Gestione Della Difesa Dagli Incendi Boschivi, pp. 71—90.

Leone, V., Lovreglio, R., Martín, M.P., Martínez, J., Vilar, L., 2009. Human factors of fire occurrence in the Mediterranean. In: Chuvieco, E. (Ed.), Earth Observation of Wildland Fires in Mediterranean Ecosystems. Springer-Verlag, Berlin Heidelberg, pp. 149—170.

Liacos, L.G., 1977. Fire and fuel management in pine forest and evergreen brushland ecosystems of Greece. Palo Alto (CA - USA) 1—5 Aug 1977. Gen. Tech. Rep. WO-3. In: Mooney, H.A., Conrad, C.E. (Eds.), Proceedings of the Symposium "Environmental Consequences of Fire and Fuel Management in Mediterranean Ecosystems". USDA Forest Service, Washington, DC, USA, pp. 289—298.

Long, M., Ripert, C., Piana, C., Jappiot, M., Lampin-Maillet, C., Ganteaume, A., Alexandrian, A., Rouch, L., 2009. Amélioration de la connaissance des causes d'incendie de forêt et mise en place d'une base de données géoréférencées. Forêt Méditerranéennet XXX (3), 221—228.

Lovén, L., Äänismaa, P., 2004. Planning of the sustainable slash-and-burn cultivation programme in Koli national park, Finland. Int. For. Fire News 30, 16—22.

Lovreglio, R., Leone, V., Giaquinto, P., Notarnicola, A., 2006. New tools for the analysis of fire causes and their motivations: the Delphi technique. For. Ecol. Manage. 234 (1), 18—33.

Lovreglio, R., Leone, V., Giaquinto, P., Notarnicola, A., 2010. Wildfire cause analysis: four case-studies in southern Italy. iFor.-Biosci. For. 3, 8—15.

Medler, M.J., 2011. Speculations about the effects of fire and lava flows on human evolution. Fire Ecol. 7 (1), 13—23.

Miralles, M., Kraus, D., Molina, D., Loureiro, C., Delogu, G., Ribet, N., Vilalta, O., 2010. Improving suppression fire capacity. Research Report 23. In: Silva, J.S., Rego, F., Fernandes, P., Rigolot, E. (Eds.), Towards Integrated Fire Management—Outcomes of the European Project Fire Paradox. European Forest Institute, Joensuu, Finland, pp. 203—215.

Montiel-Molina, C., 2011. Fire use practices and regulation in Europe: towards a fire framework directive. In: The 5th International Wildland Fire Conference Sun City, South Africa 9—13 May 2011.

Montiel, C., Costa, P., Galán, M., 2010. Overview of suppression fire policies and practices in Europe. In: Sande, J., Rego, F., Fernandes, P., Rigolot, E. (Eds.), Towards Integrated Fire Management—Outcomes of the European Project Fire Paradox. European Forest Institute, Finland, pp. 177—187. Research Report 23.

Montiel, C., Herrero, G., 2010. An overview of policies and practices related to fire ignitions at the European union level. In: Sande, J., Rego, F., Fernandes, P., Rigolot, E. (Eds.), Towards Integrated Fire Management—Outcomes of the European Project Fire Paradox. European Forest Institute, Finland, pp. 35—46. Research Report 23.

Oliveira, T., 2005. The Portuguese national plan for prevention and Protection of Forest against fires: the first step. Int. For. Fire News 33, 30—34.

Pausas, J.G., Vallejo, V.R., 1999. The role of fire in European Mediterranean ecosystems. In: Chuvieco, E. (Ed.), Remote Sensing of Large Wildfires in the European Mediterranean Basin. Springer, Berlin, pp. 3—16.

Peuch, E., 2005. Firefighting safety in France. In: Missoula, M.T., Butler, B.W., Alexander, M.E. (Eds.), Proceedings "Eighth International Wildland Fire Safety Summit", April 26—28, 2005. The International Association of WildlandFire, Hot Springs, SD, USA.

Pyne, S.J., 1995. World Fire: The Culture of Fire on Earth. Henry Holt and Co, New York, 379 p.

Pyne, S.J., 2007. Problems, paradoxes, paradigms: triangulating fire research. Int. J. Wildland Fire 16, 271—276.

Rego, F.C., Rigolot, E., Alexandrian, D., Fernandes, P., 2011. EU project fireparadox: moving towards integrated fire management. In: The 5th International Wildland Fire Conference. Sun City, South Africa, 9—13 May 2011.

Ribet, N., 2002. La maîtrise du feu un travail "en creux" pour façonner les paysages. In: Woronoff, D. (Ed.), Travail et paysages, Paris, Éditions du CTHS, Actes du 127e Congrès du CTHS «Le travail et les hommes», Nancy 15—20 avril 2002, pp. 167—198.

Rius, D., Vannière, B., Galop, D., 2009. Fire frequency and landscape management in the northwestern Pyrenean piedmont, France, since the early Neolithic (8000 cal. BP). The Holocene 19 (6), 847—859.

Roberts, N., 2004. Postglacial environmental transformation. In: Bogucki, P., Crabtree, P.J. (Eds.), Ancient Europe 8000 B.C.—A.D. 1000. Encyclopedia of the Barbarian World, vol. 1. Thomson, Gale, pp. 126—131.

Rösch, M., Ehrmann, O., Herrmann, L., Schulz, E., Bogenrieder, A., Goldammer, J.G., Page, H., Hall, M., Schier, W., 2004. Slash-and-burn experiments to reconstruct late Neolithic shifting cultivation. Int. For. Fire News 30, 70—74.

San-Miguel-Ayanz, J., Camia, A., 2010. Forest fires, in mapping the impacts of natural hazards and technological accidents in Europe: an overview of the last decade. EEA Technical Report N13/2010, 47—53.

San-Miguel-Ayanz, J., Schulte, E., Schmuck, G., Camia, A., Strobl, P., Liberta, G., Giovando, C., Boca, R., Sedano, F., Kempeneers, P., McInerney, D., Withmore, C., Santos de Oliveira, S., Rodrigues, M., Durrant, T., Corti, P., Oehler, F., Vilar, L., Amatulli, G., 2012. Chapter 5. Comprehensive monitoring of wildfires in Europe: the European forest fire information system (EFFIS). In: Tiefenbacher, John (Ed.), Earth and Planetary Sciences, Geology and Geophysics, Approaches to Managing Disaster—Assessing Hazards, Emergencies and Disaster Impacts, pp. 87—108. Publisher: InTech.

San-Miguel-Ayanz, J., Moreno, J.M., Camia, A., 2013. Analysis of large fires in European Mediterranean landscapes: lessons learned and perspectives. For. Ecol. Manage. 294, 11—22.

Sereni, E., 1981. Terra nuova e buoi rossi e altri saggi per una storia dell'agricoltura europea. Einaudi Editori, Torino, 371 p.

Steensberg, A., 1957. Some recent Danish experiments in Neolithic agriculture. Agric. Hist. Rev. 5 (2), 66—73.

Susmel, L., 1977. Ecology of systems and fire management in the Italian Mediterranean region. In: Mooney, H.A., Conrad, C.E. (Eds.), Proceedings of the Symposium "Environmental Conse-quences of Fire and Fuel Management in Mediterranean Ecosystems". USDA Forest Service, Washington, DC, USA, pp. 307—317. Palo Alto (CA - USA) 1-5 Aug 1977. Gen. Tech. Rep. WO-3.

Tedim, F., Remelgado, R., Borges, C., Carvalho, S., Martins, J., 2013. Exploring the occurrence of megafires in Portugal. Forest Ecology and Management 294, 86—96.

Vecchio, B., 1974. Il bosco negli scrittori italiani del Settecento e dell'età napoleonica. Piccola Biblioteca Einaudi, PBE 235, Torino, 276 p.

Vélez, R., 1982. Fire effects and fuel management in Mediterranean ecosystems in Spain. Gen. Tech. Rep. PSW-58. In: Conrad, C.E., Oechel, W.C. (Technical coordinators) (Eds.), Proceedings of the Symposium on Dynamics and Management of Mediterranean-type Ecosystems; 1981 June 22—26; San Diego, CA. U.S. Department of Agriculture, Forest Service, Pacific Southwest Forest and Range Experiment Station, Berkeley, CA, pp. 458—463. http://www.fs.fed.us/psw/publications/documents/psw_gtr058/psw_gtr058.pdf.

Vélez, R., 1995. Spain: the 1994 forest fire season. UN ECE/FAO Int. For. Fire Newsletter 12, 12—13.

Vélez R., 1999. The Red Books of Prevention and Coordination: A General Analysis of Forest Fire Management Policies in Spain. In:González-Cabán, A., Omi P.N. (Technical Coordinators). Proceedings of the Symposium on Fire Economics, Planning, and Policy: bottom lines, San Diego, CA, April 5—9. Albany, CA: Pacific Southwest Research Station, Forest Service, U.S. Department of Agriculture, Gen. Tech. Rep. PSW-GTR-173:171—177. http://www.fs.fed.us/psw/publications/documents/psw_gtr173/psw_gtr173.pdf.

Vélez, R., 2000. La defensa contra incendios forestales. Fundamentos y experiencias. McGraw-Hill/Interamericana de España S.A.U, Madrid, 1302 p.

Wayman, E., April 4, 2012. The Earliest Example of Hominid Fire. Smithsonian.Com. http://blogs.smithsonianmag.com/hominids/2012/04/the-earliest-example-of-hominid-fire/#ixzz2gVNw6ENM.

Xanthopoulos, G., Caballero, D., Galante, M., Alexandrian, D., Rigolot, E., Marzano, R., 2006. Forest fuels management in Europe. In: Andrews, P.L., Butler, B.W. (Eds.), Proceedings of the Conference on "Fuels Management—How to Measure Success", March 28—30, 2006, Port-land, Oregon, USA. USDA Forest Serv., Rocky Mountain Research Station, Fort Collins, pp. 29—46. CO. RMRS-P-41. 809 p.

Xanthopoulos, G., 2007a. Olympic flames. Wildfire 16 (5), 10—18.

Xanthopoulos, G., 2007b. Forest fire policy scenarios as a key element affecting the occurrence and characteristics of fire disasters. In: Proceedings of the "IV International Wildland Fire Conference", May 13—17, 2007, p. 129. Seville, Spain.

Xanthopoulos, G., 2008a. People and the mass media during the fire disaster days of 2007 in Greece. pp. 494—506. Adelaide, Australia. In: Proceedings of the International Bushfire Research Conference on "Fire, Environment and Society" of the Bushfire Cooperative Research Centre and the Australasian Fire Emergency Service Authorities Council (AFAC), September 1—3, 2008, p. 570.

Xanthopoulos, G., 2008b. Who should be responsible for forest fires? Lessons from the Greek experience. In: Proceedings of the "II International Symposium on Fire Economics, Planning and Policy: A Global View", April 19−22, 2004. USDA Forest Service, Pacific Southwest Research Station, Cordoba, Spain, pp. 189−202. PSW-GTR-208. 720 p.

Xanthopoulos, G., 2012. Evolution of the forest fire problem in Greece and mitigation measures for the future. In: Boustras, G., Boukas, N. (Eds.), Proceedings of the 1st International Conference in Safety and Crisis Management in the Construction, Tourism and SME Sectors (1st CoSaCM), Nicosia, Cyprus, June 24−28, 2011. Brown Walker Press, Boca Raton, Florida., USA, pp. 736−747.

Xanthopoulos, G., Bushey, C., Arnol, C., Caballero, D., 2012. Characteristics of wildland−urban interface areas in Mediterranean Europe, North America and Australia and differences between them. In: Boustras, G., Boukas, N. (Eds.), Proceedings of the 1st International Conference in Safety and Crisis Management in the Construction, Tourism and SME Sectors (1st CoSaCM), Nicosia, Cyprus, June 24−28, 2011. Brown Walker Press, Boca Raton, Florida, USA, pp. 702−734.

Chapter 6

Wildfires: An Australian Perspective

Petra T. Buergelt

Charles Darwin University, School of Psychological & Clinical Sciences, Darwin, Australia, University of Western Australia, Centre for Social Impact and Oceans Institute, University of Western Australia, Australia & Joint Centre for Disaster Research, Massey University, Mt Cook, Wellington, New Zealand

Ralph Smith

Department of Fire & Emergency Services, Cockburn Central, Western Australia, Australia

ABSTRACT

In Australia, wildfires are an inevitable part of the ecosystem. The wildfire risk is growing due to climate change and demographic shifts. This chapter outlines the extent and impact of wildfires, sketches historical developments and issues, examines frameworks created by legislations and institutions, and reviews inquiries, research, and education-building capacity and capability. Although Australia has made great progress in building its competence to effectively respond, current arrangements are insufficient for meeting the increasingly "wicked" problems posed by wildfires. Agreeing with Einstein that we cannot solve problems by using the same kind of thinking that created them, Australia recognizes that a new kind of thinking is required to successfully live with wildfire risk. We conclude with suggestions of how to transform the mindsets of citizens and players involved in ways that enable them to solve difficult problems and to effectively manage in a complex and rapidly changing environment.

6.1 INTRODUCTION: EXTENT AND IMPACT OF AUSTRALIAN WILDFIRES

Wildfires, or bushfires as they are more commonly referred to in Australia, are a natural and inevitable part of the Australian ecosystem (McGee and Russell, 2003). Australia's specific topographic (mountains separated by long, narrow valleys), vegetation (extensive forest coverage), and weather (hot summers with hot winds, long droughts, and lightning) features make it one of the most

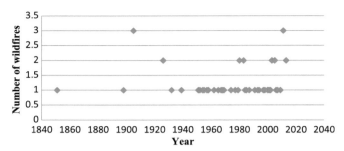

FIGURE 6.1   Australian wildfires between 1951−2013 that had significant human ecosystem impacts.

fire-prone areas in the world. Indeed, Australia's ecological system has developed symbiotically with fire. Fire influences both in situ elements (e.g., flora, soil fauna) and mobile elements (e.g., birds, mammals) of the biota (e.g., fire-stimulated flowering is common in many taxa). However, ensuring that people and the built environments they use remain safe requires managing bushfires as well as developing the capacity of residents and communities to respond to bushfires.

The risk of bushfires is increasing. Because unplanned ignitions will always occur (e.g., lightning, power-line fault, controlled burns going out of control, arson), bushfires are inevitable. Figure 6.1 provides an overview of the major Australian bushfires that impacted human ecosystems in Australia between 1951 and 2013 (Commonwealth of Australia, 2012). The most significantly destructive bushfires occurred in Victoria (21), New South Wales (19), Queensland (5), South Australia (5), Western Australia (3), and Tasmania (3) also experienced bushfires. Four bushfires effected several states at the same time: Central Western NSW and Southern Queensland, Black Friday, Ash Wednesday, and Black Christmas. Commonly, each of these fires consisted of many separate fires (e.g., Ash Wednesday fires: >100 fires forming seven regional complexes across two states) that burned for different and often long durations (e.g., Great Divide fires: contained after 69 days) and affected many different settlements (Black Saturday fires: >400 fires affected 78 communities).

Whenever flammable bushland is close to fire-vulnerable assets (e.g., people, settlements, infrastructure, and environmental values), there is a risk of bushfires having disastrous consequences. Australian bushfires substantially impacted communities and ecosystems (e.g., Dass-Brailsford, 2008; Tibbits and Whittaker, 2007). The associated ecological, household, infrastructure, and societal costs average US $1.58 billion per year (Mortimer, Bergin, and Carter, 2011). In total, Australian bushfires between 1951 and 2013 caused the death of 677 people, destroyed 9,141 homes and 3,662 buildings, led to the evacuation of 15,800 people, left 16,550 people homeless, ruined 575 farms and 130 businesses, burned 22.8 billion hectares of land, led to the loss of 537,808 lifestock, and destroyed 575 farms and 130 businesses (Table 6.1) (Commonwealth of Australia, 2012).

TABLE 6.1 A summary of Australian Fire Events and Associated Impacts

Year	Name	Deaths	Insured	Homeless/ Evacuated	Insured Costs	Homes Destroyed	Area Burnt (ha)	Livestock Died
1926	Central Western NSW & Southern QLD	60		700		69		
1939	Black Friday	71		1,000				
1967	Black Tuesday Tasmania	62			$ 610 million	1,400	264,270	20,000
1969	Lara & Melbourne Fringe	23		800			324,000	
1977	Western Districts & Streatham	4		350	$ 101 million	116	103,00	198,000
1983	Ash Wednesday	75		9,000	$ 1.9 billion	1,700	310,000	32,000
1993	North Coast, Hunter & Sydney	4			$ 215 million	206	800,000	
1994	East Coast Queensland	–		3,000	$ 215 million	23	5,000	
2001	Black Christmas	–		11,000	$ 131 million	109	753,314	7,000
2003	Canberra	4	435	5,000	$ 660 million	488	160,000	
2009	Black Saturday	173	414	41	$ 1.3 billion	2,029	450,000	
2013	Dunalley Tasmania	1		1,000		203	36,000	

Note: Cells without value means that cost did not apply or is not available.

Managing bushfires requires substantial resources. For example, extinguishing the over 100 fires burning during the 2001 Black Christmas fire through parts of New South Wales and the Australian Capital Territory for approximately 23 days required >20,000 regular and volunteer firefighters as well as 85 aircraft. Often overlooked, many firefighters and emergency personnel suffer injuries and some die. In the Great Divide Complex Bushfire in 2007, for instance, >1,400 firefighters were injured (e.g., bruises, cuts, blisters, burns, dehydration, broken limbs, and spider bites). Additional costs arise from financially assisting the recovery of affected residents and communities, agriculture, cultural heritage, and environmental assets.

6.2 AUSTRALIA'S WILDFIRE MANAGEMENT

This section sketches the historical developments regarding bushfire management in Australia and highlights issues. It discusses the particular challenge of controversies related to controlled burning.

6.2.1 Sketch of Historical Developments and Issues

For many millennia prehuman settlement, lightning has caused bushfires across Australia. Starting at least 60,000 years ago, Australian Aborigines intentionally chose to introduce and use fire (Pyne, 2006). While soil fertility, depth, type and structure, and rainfall regulate which plants grow where, this intentional use of fire might have shaped Australian biodiversity systems pre-European settlement (Gammage, 2011; Gill, 2006; Ward, 2006).

With communities becoming more static and developing significant built structures in 1814−2014, fire use and management have become less intuitive and more protective. To protect lives and artificial structures, settlers tried to reduce their risk by modifying fuel levels through intentionally initiating planned burning. However, initial European settlers lacked local knowledge. Instead, their management actions were based on beliefs derived from practices based on European climate and agriculture. They brought livestock that chewed the vegetation; they cleared the bush to plough, sow seeds, and grow crops and they fenced off extensive forest reserves and excluded fire from being introduced. Their use of fire regimes cleared the native endemic vegetation and restructured fuels. In many locations, European practices led to an extinction of Aboriginal practices (Pyne, 2006).

Through other activities, settlers unintentionally increased the risk of bushfires due to the lack of local knowledge of the ecosystems. Logging, for instance, changed the forest structure toward producing a heavier fuel load. Originally, forests of mature trees completely controlled and suppressed ground flora, resulting in low levels of fuel. This low bushfire fuel load, in turn, meant that fires that were lit or caused by lightning burned slowly with a low intensity. In contrast, logging, which was used to create timber for housing and other development, created large gaps in the canopy that allowed sunlight to get to the ground flora resulting in more vigorously growing ground scrub flora and understorey species. This growth combined with the post logging debris, if not reduced, significantly increased fuels, and thus resulted in fast moving and high intensity fires (Beggs, 1971). Further research has indicated that frequent burning can increase species diversity in the understorey, but decrease species diversity in the overstorey (Marsden-Smedley, 2009).

By the nineteenth century, particularly late in the century, Australian foresters had determined that managed fire did belong in the bush environment, and an innovation in land management was born (Pyne, 2006). Western Australia Forests Department Conservator Kessell identified that fire control was possible, but complete fire prevention was impossible to achieve in the jarrah belt in the southwest of Western Australia. Later prescribed burning techniques using aerial ignition via fixed- and rotary-winged aircraft enabled large areas to be effectively and efficiently burned under prescribed conditions.

During the post-1960 era, forest departments across Australia and within the Commonwealth Government, and Commonwealth Scientific and Industrial Research Organization (CSIRO) did an excellent job in developing fire-spread models for forest fuels (e.g., Forest Fire Danger Meter, CSIRO Grassland Fire Spread Meter, and CSIRO Fire-Spread Meter for Northern Australia) (Rodger, 1961). Whilst these fire-spread meters are very good, particularly for prescribed burning, they are all known to underestimate the head fire rate of spread in summer during extreme fire events because they were developed using single-point ignitions rather than lines of fire and/or low to moderate forest fire danger. The "Project Vesta" research and the "Semiarid mallee heath fire spread model" partially address this issue but had a mixed uptake, possibly due to their being to much of complex (e.g., regarding fuel load and structure) (Gould et al., 2007; Cruz, 2010).

Fire and land management agencies and private land managers (e.g., farmers, graziers, and pastoralists) utilize the published fire spread models when determining the most appropriate bushfire management options. Many of the fire and land management agencies have interpreted these models and applied parameters for their application to meet the needs of the community. What is still lacking is a national standard approach to fuel load assessment and fire spread meters that are appropriate for prescribed burning, and also for summer fire spread for all major vegetation types.

During the 1970 and 1980 landscape hazard reduction and ecological burning in the pastoral regions and forests of the southwest of Western Australia were at a historical high. In the post-1980 era, the role of forest departments across Australia changed to a point where there are now no State Forest Departments with significant and integrated land-management responsibilities. That land management role has been effectively moved to conservation-based land management agencies. This change in forest management has resulted in a reduction in crews working daily in the bush, and a reduction in the availability of skilled timber harvesting bush workers and heavy production machinery that was previously available for use in fire suppression.

Since the 1990s, urban sprawl into the rural—urban, periurban, or wildland—urban (bush/forest-settlement) interface has increased dramatically, while landscape burning in the urban—forest interface of the southwest of Western Australia has declined. This combination exposes urban fringe areas to a significantly increased bushfire risk. Virtually every inquiry conducted after major Australian bushfires since the early 1960s has identified the lack of landscape fuel management as a major contributing factor to fires becoming so destructive (e.g., 2009 Victorian Wildfires Royal Commission, 2010; Bateman, 1984; Esplin et al., 2003; Gill and Moore, 1997; Keelty, 2011; Lewis et al., 1994). This conclusion is supported by recent research that shows for different landscapes that the proportion and impact of destructive, late dry season bushfires are reduced when early dry season burning is undertaken (personal communication with Adrian Allen from Landgate in 2012). The other consistent outcome of the significant inquiries is that additional finance needs to be made available in the long term to ensure that the fuels are maintained at an appropriate level in perpetuity.

The combination of heavy fuel loads and inappropriate of house construction standards are exposing the community to significant wildfire risks. Homes not constructed to appropriate Bushfire Attack Level Standard (Australian Standards, 2009) and excessive fuel loads too close to buildings and within the landscape results in a large number of destroyed and damaged homes (Smith, 2011a,b). This risk factor could relatively easily be rectified through developing, implementing, and reinforcing appropriate building construction standards. The building-construction standards need to take into account the environment. For instance, when a building is associated with a pasture-grass type of fuel, the level of construction protection required is significantly less than when built immediately adjacent to a forest environment. In both environments, homes need to be built in ways that ensures that embers do not enter, direct flame contact does not occur, and radiant heat attack on the building falls short of the design standard for this hazard consequence.

However, recent inquiries have identified that buildings being placed in the rural—urban interface are not being constructed to the appropriate standard. This is occurring even though there are general national (e.g., "Australian Standard 3959—Construction of Buildings in Bushfire-prone Areas") and specific state guidelines in place, as well as material available to assist residents to determine building requirements.

6.2.2 Controversies Regarding Prescribed Burning

Rainfall in Australia is very variable, with significant rainfall in some years and virtually none in others. The seasonality and volume of rainfall influence levels of vegetation growth (Latz, 2004). From a bushfire management perspective this means there will be periods when fuel loads are conducive for large, unmanaged bushfires, unless the fire fuels are managed. However, in the rural—urban—forest interface, there is often conflict between the management of the vegetation to reduce the fire hazard and the management of the vegetation for conservation, and the maintenance of the amenity or lifestyle value of bush or forest.

This creates a need to reconcile the protection of communities and assets in those living close to forests with ecosystem protection. For some, such as those community members who espouse environmental values, this can be challenging if people put environmental protection ahead of social protection. Accordingly, this can, in some instances, require a compromise between community protection and biodiversity requirements. Effectively reconciling the protection needs of built and natural environments, which are not mutually exclusive, requires appropriate knowledge, planning, and plan implementation.

The management and reduction of fuel loads make early fire suppression a viable option. Managing fuel loads increases the ability of firefighters to suppress bushfire and to protect the community, and thus reduces social disruption and its associated costs. As the ability of firefighters to directly attack the fire declines, the intensity and the area burned by the bushfire increase, and the level of impact on communities (e.g., loss of lives, loss and damage of houses, and displacement) and on natural environment increases.

The intensity of a bushfire is determined by the rate of spread of the fire, the heat yield of the burning vegetation, and the amount of fuel available to be burned. The only component of these inputs into the formulae that can be managed is in the amount of fuel available. The prescribed use of fire is potentially the least intrusive and the ecologically friendliest and most cost-effective method of protecting built structures and modified landscapes. However, using fire appropriately is one of the most complex and divisive issues within communities.

Fire management is complex. No single fire regime is optimal for all biota. When factoring the requirement to protect human life and community values, the situation becomes significantly more complex and one that can create conflict (Paton et al., 2008). This conflict may result in the compromising of some biodiversity values or precluding some areas from development. The acute impacts on biota and postfire recovery time are determined by the seasonality, frequency, scale, and intensity of the fire (Burrows and Wardell—Johnson, 2003). Whether prescribed fire is introduced or excluded needs to be based on the land use priorities for the particular site.

Effective and appropriate management of the fuel load does increase the protection afforded communities and the natural environment. Fire has only an adverse impact on natural environments, including biodiversity, if it is too frequent, too intensive, too long excluded, or occurs in the same season too frequently (Burrows and Wardell—Johnson, 2003). Whether the impact is permanent or transitional is determined by the vegetation type, structure, adaption to fire, fire frequency, intensity, and season. A properly managed fuel reduction program will cause far less damage to life and property, biodiversity, and ecological processes than uncontrolled and intense bushfires (Adams and Attiwell, 2011). Hence, it is crucial to find the right balance: too much fire can alter the composition and structure of the natural ecosystem; too little can predispose the natural ecosystem to larger and more severe bushfires (Waller, 2011). Factors such as the spread direction add further complexity.

Fire-spread direction and speed patterns are extremely variable, depending as they are on the combination of vegetation type and condition, weather conditions, and the number of fires joining together (Burrows, 1984; Gould et al., 2003). Research is contributing to our understanding of these interactions. For instance, when two fires join (junction zone), the rate of spread will increase by around twice the previous rate of spread. Junction zones of fires with 100- to 200-m fire lines will increase their rate of spread between 2 and 2.5 times (Gould et al., 2003). Therefore, to minimize the fire behavior on the vegetation, it is essential to have junction zones occurring on a falling hazard (generally late afternoon when temperatures drop and humidity rises) and with shorter fire lines.

The frequency of planned burning and the nature of the spatial mosaic to maintain fuel loads at acceptable levels are determined by the potential human and natural values at risk, rate of fuel accumulation, the nature of the fuel, the climate of the region, and the fire detection and suppression capability. A significant complicating factor is that for many ecosystems the rate of fuel accumulation is not known and there is no specific fire-spread meter available. However, there are now processes available to identify and minimize the potential adverse impact on biodiversity (Burrows and Wardell-Johnson, 2003). For example, burning in spinifex in desert regions every 2–4 years promotes perennial grasslands (Enright et al., 2012; Gammage, 2011).

The expansion of the environmental protection legislation into the restrictions of clearing of native vegetation and the inclusion in that legislation of fire as a clearing mechanism have introduced statutory complexity into prescribed burning. This legislation led to land owners or holders being more reluctant to undertake prescribed burning. The absence of a State fuel-load strategy or criteria has added further complexity. The question of when the fuel load is excessive is very subjective without such specific criteria, particularly when compared to the impact and constraints of the environmental protection legislation. The difficulty with the application of any prescribed burning criteria is to know what the slowest maturing, fire-sensitive, obligate-seeding species are within that vegetation complex or region and the competing needs of protecting the community and the community-built assets.

With the recent changes to the structure of Government Agencies across Australia, there has been a movement to provide more statutory power in regard to fire management to responding agencies, rather than to conservation and/or land-management agencies. These fire and conservation agencies are undertaking prescribed burning on behalf of, or in partnership with, a range of land owners/managers. This burning is undertaken in highly diverse and complex vegetation locations.

In Victoria, for instance, fire fighting is divided between three agencies responsible for different areas (Waller). The Department of Sustainability and Environment is responsible for suppressing fires on public land. The Country Fire Authority is responsible for fighting fires on private land in the country. The Metropolitan Fire Brigade is responsible for fighting fires metropolitan Melbourne. All three use the Australasian Interservice Incident Management System and have long-standing arrangements to jointly prepare for and fight bushfires.

Across Australia, local governments and local residents, including pastoralists and farmers, seek the assistance of government agencies to augment their own bushfire-management experience and knowledge. Local communities seeks this expert assistance, and often request that the response is provided quickly and effectively to protect that community from a potential bushfire. To fulfill this need, the procedures that Government Agencies implement in regard to prescribed burning planning and implementation must be able to cater for both the regular, annual well-planned, and implemented burning program and the short time frame, quick response but equally well-planned and implemented burning program to meet community expectations.

Since the 1980s, the level and type of equipment used in bushfire management has improved significantly. There has been a national expansion in the use of aerial suppression, whether by rotary or fixed wing aircraft. These appliances are expensive to utilize and have known fireline intensity effectiveness restrictions. Community beliefs that a bushfire is not significant unless aerial fire suppression is engaged and applied represents an unsustainable community expectation and one that will result in significant financial costs for potentially little operational benefit if it is permitted to remain. Climate change pressure will potentially amplify these problems (e.g., rainfall decline resulting in less spring weather suitable for prescribed burning). Recognition of the significance of these issues has promoted several legislative responses.

6.3 FRAMEWORK: LEGISLATION AND KEY INSTITUTIONS

In Australia, the states/territories hold the primary responsibility for bushfire and land management within their jurisdictions (as described in the Australian Constitution). In regard to disaster and emergency management, the Australian Government provides leadership through various whole of Government Ministerial Councils (Federal, State, and Territory Ministers), frameworks, and strategies to guide the development of operational capacity of the

Commonwealth, states, and territories to respond to and recover from di-sasters. In this section, we will first outline the national frameworks and strategies followed by discussing how the state/territories implement them. It will become apparent that while in 2004–2014, an extensive body of sound frameworks and strategies has been developed at the national level, the translation of these into action rely on goodwill and at the state and local levels has been largely ineffective due to structural, practical, and human barriers.

6.3.1 National Level: A Sound Body of Frameworks and Strategies

The Australian constitution provides clear guidance as to the approach and responsibilities regarding emergency management in general and bushfire in particular. The major general emergency-management documents include the National Disaster Resilience Framework; the National Strategy for Disaster Resilience; the National Emergency Risk Assessment Guidelines; and the National Action Plan for the Attraction, Support, and Retention of Emergency-Management Volunteers. These general documents legislations are compli-mented by legislation specific to bushfires: the National Bushfire Policy Statement, the National Work Plan to Reduce Bushfire Arson, and the National Strategy for the Prevention of Bushfire Arson.

The legislation seek to strengthen capabilities of individuals, communities, organizations, and businesses to respond to extreme natural events by creating a comprehensive, coordinated, and cooperative whole-of-nation, resilience based, all-hazard PPRR prevention, preparedness, response and recovery (PPRR) framework that involves all sectors at all levels (Commonwealth of Australia, 2013a,b). They provide high-level, strategic direction, guidance, and principles for the emergency-management sector to develop disaster-resilient communities across Australia that function well under stress, successfully adapt, are self-reliant, and have appropriate social capacity.

The objective is to create enduring partnerships, shared understanding and responsibility, and sustained behavioral change. They also aim at creating comprehensive and integrated understanding of emergency risks, and improving the national evidence base on risks in Australia (Commonwealth of Australia, 2013c). Similar to Canada and the United States, emphasis has shifted from responding to preventing bushfires, to proactively reducing the risk they pose, and simultaneously facilitating engaging communities in preparing (Waller, 2011). While the policy statements promote overarching national strategic ob-jectives and shared principles, they allow sufficient flexibility for jurisdictions to determine their own bushfire-management policies.

In 1993, the Australian Assembly of Rural Fire Authorities and the Australian Assembly of Fire Authorities joined forces to reduce duplication, pool resources, and address common problems. They created the Australasian Fire and Emergency Service Authorities Council (AFAC)—a peak body for

Australasia fire, land management, and emergency management (http://www.afac.com.au/home). Members include public and private sector organizations. Its large scale and scope enable the AFAC to tackle big and complex problems that require collaborative, synergistic efforts across the industry.

6.3.2 State and Local Levels: The Implementation Challenges

The national emergency and bushfire-management framework is implemented by each state/territory by a complex system of various state government departments and organizations. In Western Australia, for instance, the Department of Fire and Emergency Services, Department of Parks and Wildlife, and the State Emergency-Management Committee are the main players involved in managing bushfires. However, due to the complex nature of bushfire preparedness and response, many other state government departments (e.g., Department of Health, Department of Education, Department Child Protection, and Family Support, Department of Housing), associations and peak bodies (e.g., Western Australian Local Government Association), NGOs and service providers (e.g., Red Cross), local governments (e.g., Shire Emergency-Management Committee), social and natural researchers, public and private education providers, media, and businesses are and need to be involved (Buergelt & Paton, 2014).

The issue is that, in practice, the degree of collaboration and coordination among and between these key players at all scales and across states/territories is not as effective as it could be due to differences in perspectives and approaches, lack of superordinate leadership, and effective organizational structures connecting them, the compartmentalization of government departments, frequent restructuring, staff being insufficiently trained in emergency management and collaboration, staff being increasingly stretched and overwhelmed, and high staff turnover.

Narrow and inconsistent approaches to collaboration and coordination seriously undermine the effectiveness of bushfire management. It results in a lack of accountability, duplication of efforts, and lack of pooling resources. It weakens the ability of agencies to conceptualize the emergency-management system as a whole, and reduces the quality of the interconnections between the different parts and players. This, in turn, compromises the ability to create strategic long-term plans, integrative approaches, and synergistic and creative solutions that address those aspects of the whole system that afford the greatest impact on risk management. The end result is that the frameworks, principles, and actions provided by the federal leadership are not translated into effective actions at the ground.

6.4 BUILDING CAPACITY AND CAPABILITY: INQUIRIES, RESEARCH, EDUCATION, AND TRAINING

The building of capacity and capability in bushfire and emergency management is accomplished via public inquiries, research, education provided by government agencies, and training offered by public and private education providers. What follows is a discussion of these aspects. It will become obvious that while Australia has put a lot of resources into assessing what did not work in previous bushfires, it has not, in general, been effectively learning from and implementing lessons learned from previous fire events. We also argue that the research, and the education and training of emergency management, is not as effective as it could be.

6.4.1 Inquiries into Bushfires: Identifying and Learning Lessons

Several bushfires have prompted inquiries, conducted by different agencies, into many different aspects of bushfire management at federal, state/territory, and/or local scales. Besides facilitating learning lessons from fires, inquiries also serve to ensure justice, that the voices of affected community members are heard, and to ensure that the memory of fires is not lost from public consciousness (Teague et al., 2010). Data are collected through reviewing previous reports and research, conducting open hearings of the various players involved, community meetings and interviews, web streaming, and public submissions.

The most prominent recent inquiries include the COAG Council of Australian Governments (COAG) National Bushfire Inquiry and Response (Ellis et al., 2004), the Nairn Inquiry into Australian Bushfires (House of Representatives Select Committee into the recent Australian bushfires, 2003), the 2009 Victorian Bushfires Royal Commission Report (Teague et al., 2010), the 2011 Inquiry into the Perth Hills Bushfires (Keelty, 2011), and the 2011 Special Inquiry into the Margaret River Bushfire (Keelty, 2012).

Aspects investigated by these inquiries comprised, for instance, the effectiveness and appropriateness of Australia's disaster management approach, the reasons for prescribed burns escaping, the causes of and responses to bushfires, the preparation and planning, measures taken in relation to utilities, the ability of agencies to collect data and inform government and community about risk and its mitigation, the governance around critical decision making, resource allocation, communication, and effective measures.

These inquiries influenced public perspectives, policies, and guidelines but to a lesser extent actual practices. For example, the 2003 COAG Report on Natural Disasters in Australia recommended a historical shift in focus from response and reaction toward cost-effective, evidence-based disaster anticipation and mitigation, and a systematic and widespread national process of disaster risk assessments (Ellis et al., 2004). The 2009 Victorian Bushfire Royal Commission Report shifted fire management in Australia to fire prevention based on land and fuel management, management on landscape scale, adaptive management based on high-quality, sustained research, and management that engages the community to ensure full public involvement (Waller, 2011).

The inquiries identified many issues and made a wealth of recommendations. However, they reiterated many issues and recommendations identified in previous reviews (Ellis et al., 2004; Keelty, 2011). Critical issues identified included community complacency before every major fire event, consistently poor resourcing, concerns about the need for protective burning, the roles and responsibilities of local government, and the role and contributions of the insurance industry. The key common recommendations centered around risk reduction activities (e.g., school- and community-based education and awareness, clearing of fuel around buildings, track access and maintenance, and fuel reduction). All inquiries emphasized the value of volunteers, the importance of communication and telecommunications, and the significance of local knowledge, and engaging people who possess local knowledge.

The reiteration of issues and recommendations suggest that lessons of history were not and are not being learned, increasing the likelihood that past mistakes will, and are being repeated. Consequently, the issue is how to effectively learn from previous enquiries and ensure that these lessons are passed in the fabric and legislative and community life. The consequent need for a proactive response to the issues identified in the inquiries has led to the development of several research programs. It is to a discussion of this scientific research that this chapter turns now.

6.4.2 Advancing, Integrating, and Transferring Knowledge: Scientific Research

As is the case with government agencies, the research landscape in Australia is complex and fragmented. The CSIRO; various universities; Commonwealth, State, and Territory Government Departments; the Bushfire Cooperative Research Centre; and the new Bushfire and Natural Hazards Cooperative Research Centre (BNHCRC) are the major research players. However, sprinkled over Australia, there exist at different universities various research institutes and centers that conduct research into the natural and social science aspects of bushfire and emergency management. In addition, there are many researchers at many different universities who conduct their specific research in relative isolation.

CSIRO, Australia's national science agency, has, since the 1983 Ash Wednesday fires, conducted research into diverse aspects of all Australian bushfires (http://www.csiro.au/Outcomes/Environment/Bushfires.aspx). CSIRO bushfire researchers covered, for example, understanding and predicting bushfire behavior, the impact of bushfires on infrastructure, the ecological responses to fire, the impact of climate change on bushfire risk, and pollutants and greenhouse gases resulting from bushfires.

Research has also addressed weather warnings, fire location information, firefighter training, predicting fire behavior, and informing fire safety policy, and led to the development of tools, methods, guides, and training materials used by fire emergency services agencies across Australia. Together with other bushfire experts, CSIRO are providing vital information and improve technologies and strategies regarding predicting bushfire behavior and managing bushfires.

In response to a series of serious bushfires, the national Bushfire CRC was formed in 2003 (http://www.bushfirecrc.com/). The Bushfire CRC sought to create an economically and ecologically sustainable, bushfire-management strategy by providing a research framework that enhanced the effectiveness of bushfire-management agencies and increased the self-sufficiency of communities. To this end, the Bushfire CRC created a multidisciplinary, multi-institutional research environment that concentrated on addressing the needs of end users, and developed strong and lasting collaborative links between researchers and private and public sector end users.

In 2013, recognizing the need to further step up Australia's resilience and the considerable impact of other natural hazards, the Australian Prime Minister contributed about $47 million over eight years to establishing a new research center (http://www.bnhcrc.com.au/). The objective of the resulting BNHCRC is to expand the Bushfire CRC research into other natural hazards. The BNHCRC will focus on creating a long-term research base for conducting coordinated, interdisciplinary, end-user inspired, and high-quality applied research, together with communities that addresses strategic issues of national significance and that is driven by the needs of end-users (Call for Proposals, 2013). It aims at contributing to the development of cohesive, evidence-based policies, strategies, and tools, and to support emergency services in all aspects of risk management.

Realizing that the problems faced are 'wicked' problems that are complex, characterized by a wealth of interconnected relationships and dependencies, the BNHCRC identified and scoped five interconnected problem areas, and encourage policy makers, practitioners, and researchers to think in new ways and to develop research that takes into account, and addresses the complex and interdependent nature of disaster issues. Figure 6.2 depicts that the areas BNHCRC research is going to work on across all Australia relevant hazards including bushfires in the next eight years.

Independently, some Australian universities created research programs, centers, or networks to bring together research expertise from across their universities in areas related to bushfire and emergency management. These include the Centre for Risk and Community Safety and the Emergency Management and Disaster Research Networks at the RMIT University, the National Climate Change Adaptation Research Facility, the Natural Disaster Management Research Initiative at the University of Melbourne, and the Appleton Institute at Central Queensland University.

These centers conduct research on a wide variety of disaster-related topics, but focus primarily on bushfires. Current projects include reviews of bushfires and bushfire research, probability of fire ignition and escalation, challenges

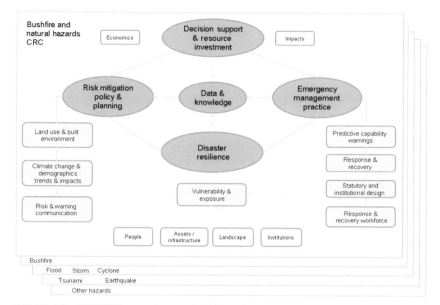

FIGURE 6.2   BNHCRC research areas. *Bushfire and Natural Hazards CRC: Call for Proposals, 2013.*

and opportunities for the Victorian emergency-management sector, review of community safety during recent bushfire events, sharing responsibility across policy sectors, building best practice in child-centered disaster risk reduction, children's knowledge of vulnerability and resilience to bushfires, effect of animals on human behavior in natural disasters, operational readiness of volunteer firefighters and building resilience through nontraditional emergency management volunteering.

6.4.3 Building Professional Capabilities: Training of Professional and Volunteer Talent

In Australia, the bushfire and emergency-management capabilities of professional staff and volunteers are built up through courses of different lengths and at different levels offered by academic, public, and private providers. However, the effectiveness of this training is compromised by the diverse players operating largely in isolation from one another rather than collaborating and coordinating to leverage their impact.

Increasingly, emergency management has become recognized as a distinct field of study that is required for progression to senior and executive levels of management in statutory emergency services; community and health services; the military; and local, state, and federal government. Consequently, emergency management is to a greater extent acknowledged as an important area to teach at universities to enhance leadership and management capabilities in the emergency-management sector.

Some universities (e.g., Australia National University, University of South Australia, Charles Darwin University, and James Cook University) offer graduate and postgraduate courses and units in emergency management and related fields (e.g., health and safety). Commonly, the courses and units are offered by a wide variety of schools which collaborate to deliver the courses. For instance, Australia National University's Research School of Earth Sciences, Crawford School of Public Policy, and Joint Colleges of Science offer together a two-year Master of Natural Hazards and Disasters. Courses are also offered by several universities collaborating as in the case of the Torrens Resilience Institute at the University of Adelaide, which pools the expertise of academics from the University of Adelaide, Cranfield University, Flinders University, and the University of South Australia.

The Australian Emergency Management Institute (AEMI) has played a significant role in building knowledge and skills capacities in the Australian emergency-management sector since 1956 (http://www.em.gov.au/Education/Pages/default.aspx). AEMI is responsible for developing community and emergency-management sector understanding and capability, awareness, and resilience; strategic consideration of emerging emergency-management capability issues of national importance; lessons learned; applied research; leadership; and practitioner competency to a consistent level across Australia.

Private training organizations also offer emergency-management courses to governments, NGOs, and volunteer agencies. Registered Training Organizations provide certificate, diploma, and advanced diploma courses in risk and emergency management, public safety, fire-fighting management, incident management, community safety, and life skills. Moreover, some teach relevant generic skills such as leadership and communication. They also offer opportunities to develop internal learning systems for companies, accredited training in the workplace, specialist skill set training, exercises/simulations, and e-learning capabilities. Some training organizations developed their programs in consultation with industry and manufacturers. Examples of private training organizations are the Australian College of Community Safety, International Academy of Law Enforcement and Security, and the Risk, Response, and Rescue. The professional development opportunities afforded by these endeavors are further supported by various conferences.

Australia hosts two conferences where practitioners and researchers share knowledge and experiences: the "AFAC Conference" and "Earth: Fire and Rain." The "AFAC Conference" is Australasia's largest and most important emergency management and public safety conference and trade expo. The conference is staged annually by the AFAC in cooperation with the Bushfire CRC. The presentations are given by practitioners and researchers predominately from AFAC member organizations and the Bushfire CRC.

"Earth: Fire & Rain" is the Australian and New Zealand Disaster and Emergency Management Conference. This three-day international conference is a joint initiative of three not-for-profit organizations: the Australian Institute of Emergency Services, the Australian and New Zealand Mental Health Association Inc., and the Association for Sustainability in Business Inc. It brings together key stakeholders and provides a forum to examine issues surrounding natural and artificial disasters.

6.5 CONCLUSION: WAYS FORWARD

Taken together, while Australia has made great progress in building its capacity to respond effectively to extreme natural events like bushfires, the current arrangements are insufficient for meeting the challenges ahead (BNHCRC Call for Proposals, 2013). We live in an age of uncertainty, insecurity, and wicked problems such as extreme natural events such as bushfires. Bushfires will, as a consequence of climate change, increase unless innovative interventions will intercept this trend. Bushfires are likely to result in more serious impacts. Demographic shifts (e.g., immigration, see and tree change of baby boomers) are leading to a growing number of people, including those traditionally identified as highly vulnerable (e.g., elderly persons), living in disaster prone areas. More players are participating in preparing, responding, and recovering. A complex set of interacting psychological, historical, natural and physical, social, spiritual/religious, technological, economic, and political factors are involved, influencing each other over time (Buergelt and Paton, in press).

At the same time, current policies and practices are not only progressively struggling and proving to be inadequate for meeting the challenges ahead, but often increase the risk and become unsustainable (BNHCRC Call for Proposals, 2013). Human resources are increasingly stretched, inadequate trained and managed for the task of addressing these situations, and disaggregated. Education has not made much difference (Lindell, Arlikatti & Prater, 2009). The new solution, community engagement, is not achieving the desired results either. This is not because community engagement does not work but because stakeholders have insufficient capabilities to truly and effectively engage with each other. The result is failure to facilitate effective, sustained, and coordinated action at person, community, and societal levels.

Einstein, in emphasizing that "insanity is doing the same thing over and over again and expecting different results" and that "we can't solve problems by using the same kind of thinking we used when we created them," pointed to the need to transform[1] people's mindset. Wicked problems can only be solved, and people can only cope with rapid change if the majority of people in organizations are learning fast, understanding complexity, synthesizing seemingly separate

---

1. Changing results into different formats or types while maintaining the substance; transforming results in an entirely different substance.

pockets of knowledge, and are able to apply this new knowledge to create innovative solutions and products/services (Berkes et al., 2003). That, in turn, requires unprecedented levels of cooperation, and high levels of analytical, critical, clear, and creative thinking. The key to success in the twenty-first century societies will be the knowledge of how to navigate through rapid change, and to inspire and lead people and groups through this change. It is not only about possessing the knowledge but also about mastering the passing of that knowledge on to others and thus facilitating the evolution of change (Berkes et al., 2003).

Based on the discussion in this chapter, we suggest some ways forward that more or less transcend current thinking. To fill knowledge gaps in bushfire management, a national standard approach for determining bushfire fuel levels needs to be developed. This approach needs to be further enhanced using new, nondestructive methods potentially linked to new technology. There is also a need to expand fire spread models so that they accurately predict the fire rate spread from single point ignitions, line of fire and within the full range of fire management from low intensity prescribed burns, through to extreme fire behavior events for all major vegetation types. Additionally, the fire-spread models must work at the landscape scale and microscale or within the suburb scale.

A house flammability model needs to be linked into the microscale or within suburb scale, fire-spread meter. To be useful, a prescribed burning framework will need to be sophisticated enough to ensure that the community protection needs and expectations are accomplished and that statutory and administrative requirements are met. At the same time, it needs to be suffi-ciently simple, flexible, and practical to be adoptable by local government, residents, and other government departments.

To increase the capability and performance of the major governmental, NGO, industry and academic players involved in bushfire and emergency management, they need to be mapped at the federal, state/territory, and local scales. The mapping needs to include mission, vision, roles, structures, staff, partners, tasks, programs, what works and does not work, and needs and solutions to get a sense of who is doing what, who is interacting with whom, what works, and what does not work. Gaining this overview enables identifying duplications, gaps, and opportunities for pooling resources and collaborating. To be useful, the map would need to be a living document to reflect constant changes. Based on that map, effective structures that facilitate collaboration could be created in ways that are sustainable. Effective collaboration and coordination can also provide the foundation for developing innovative and creative new possibilities for solving issues that we cannot conceive of from our limited thinking now.

To utilize the potential of collaborating, staff needs to be trained in developing the mindset and skills necessary for collaborating, and new structures that facilitate collaboration need to be created (e.g., sociocracy). Cutting-edge learning and teaching approaches (e.g., accelerated and transformative learning, quantum teaching) and communication technologies (e.g., webinars, google hangout) could be used for increasing the effectiveness of education and training of professionals, volunteers, and communities.

Interestingly, the recommendations that have been forthcoming from all inquiries are not crossreferenced with the scientific literature. Conversely, researchers commonly neglect the findings of inquiry reports. Bridging this rift between practice and science by both camps requires integrating both bodies of knowledge in ways that ensure that the whole is greater than the sum of the bushfire knowledge and practice parts.

While the Bushfire CRC greatly enhanced collaboration between and among end users, government and researchers, limited funding meant that only selected researchers and research groups could be included in the network. A peak body that unites and empowers "all" Australian disaster researchers; creates an inclusive, long-term, and coordinated national research agenda; and fosters collaboration and coordination among researchers and between re-searchers and the other key players is missing. Such a disaster researchers peak body would greatly enhance the comprehensiveness, effectiveness, usefulness, and applicability of the research.

A consistent, coherent, coordinated, and strategic framework and network that covers all education providers (e.g., government, industry, and academia), and works closely together with relevant government and industry bodies seems also missing. A peak body for bushfire and emergency-management training could fill this gap and ensure that the necessary professional and volunteer ca-pabilities at the different scales are effectively build.

Finally, but most importantly, to not only survive but also possibly thrive in the face of living with bushfire risk and to create safe and thriving commu-nities, it is necessary to equip citizens with matching physical, mental, emotional, and spiritual capabilities. Given that using traditional informative education has not achieved the desired results (Lindell, Arlikatti & Prater, 2009), it might be worth testing alternative education technologies and tech-niques that apply scientific breakthroughs to make learning astoundingly faster, increase memory, inspire learning, and motivate actions (e.g., trans-formative education, experiential learning, accelerated learning, and quantum teaching). Likewise, it might be useful to utilize various techniques that have been used for millennia to positively affect physical, mental, emotional, and spiritual well-being, and that are increasingly proven to facilitate these effects by scientific research (e.g., yoga and meditation).

REFERENCES

Adams, M., Attiwell, P., 2011. Burning Issues Sustainability and Management of Australia's Southern Forests. CSIRO and Bushfire CRC, Melbourne.

Bateman, T.H., 1984. Final Report of the Select Committee of the Legislative Assembly Appointed to Inquire into Bush Fires in Western Australia. Government of Western Australia, Perth.

Beggs, B.J., 1971. Forestry in Western Australia. Forest Department, Perth.

Berkes, F., Colding, J., Folke, C., 2003. Navigating Social—Ecological Systems: Building Resilience for Complexity and Change. Cambridge University Press, Cambridge.

Buergelt, P.T., Paton, D., 2014. An ecological all-hazard inter-disciplinary risk management and adaptation Model. Hum. Ecol. 42 (4), 591—603.

Burrows, N.D., 1984. Describing Forest Fires in Western Australia: A Guide for Fire Managers. Technical paper No 9. Forests Department, Perth.

Burrows, N., Wardell—Johnson, G., 2003. Fire and plant interactions in forested ecosystems of south-west Western Australia. In: Abbott, I., Burrows, N. (Eds.), Fire in ecosystems of southwest Western Australia: Impacts and management. Symposium proceedings, 16—18. April 2003, vol.1, pp. 225—268. Perth, Australia.

Commonwealth of Australia, 2012. Australian Emergency Management Hub: Bushfire. Retrieved from: http://www.emknowledge.gov.au/category/?id=1.

Commonwealth of Australia, 2013a. National disaster resilience framework. Retrieved from: http://www.em.gov.au/Publications/Program%20publications/Pages/NationalDisasterResilience Framework.aspx.

Commonwealth of Australia, 2013b. National strategy for disaster resilience. Retrieved from: http://www.em.gov.au/Publications/Program%20publications/Pages/NationalStrategyforDisaster Resilience.aspx.

Commonwealth of Australia, 2013c. National emergency risk assessment guidelines. Retrieved from: http://www.em.gov.au/Publications/Program%20publications/Pages/NationalEmergencyRisk AssessmentGuidelines.aspx.

Cruz, M., 2010. Quick Guide for Fire Behaviour Predictions in Semi-Arid Mallee-Heath. CSIRO, Canberra.

Dass-Brailsford, P., 2008. After the storm: recognition, recovery and reconstruction. Prof. Psychol. Res. Pract. 39, 24—30.

Ellis, S., Kanowski, P., Whelan, R., 2004. National Inquiry on Bushfire Mitigation and Management. Commonwealth of Australia, Canberra.

Enright, N.J., Keith, D.A., Clarke, M.F., Miller, B.P., 2012. Flammable Australia fire regimes, biodiversity and ecosystems in a changing world. In: Bradstock, R., Gill, A.M., Williams, R.J. (Eds.), Flammable Australia Fire Regimes, Biodiversity and Ecosystems in a Changing World (217). CSIRO Publishing, Collingwood.

Esplin, B., Gill, M., Enright, N., 2003. Report of the Inquiry into the 2002—2003 Victorian Bushfires. Government of Victoria, Melbourne.

Gammage, B., 2011. The Biggest Estate on Earth: How Aboriginals Made Australia. Allen & Unwin, Crows Nest.

Gill, A.M., Moore, P.H.R., 1997. Contemporary Fire Regimes in the Forests of Southwestern Australia. Report to Environment Australia. Centre for Plant Biodiversity Research and CSIRO Plant Industry, Canberra.

Gill, A.M., 2006. Dating Fires from Balga Stems: The Controversy over Fire Histories Determined from South-Western Australia Xanthorrhoea. Report to the Western Australian Department of Environment and Conservation. Western Australian Department of Environment and Conservation, Perth.

Gould, J.S., McCaw, W.L., Cheney, N.P., Ellis, P.F., Knight, I.K., Sullivan, A.L., 2007. Project Vesta—Fire in Dry Eucalypt Forest: Fuel Structure, Fuel Dynamics and Fire Behaviour. Ensis-CSIRO, Canberra ACT and Department of Environment and Conservation, Perth.

Gould, J.S., Cheney, N.P., McCaw, L., Cheney, S., 2003. Effects of Head Fire Shape and Size on Forest Fire Rate of Spread. Paper presented at the AFAC Bushland Fire Conference, Sydney, Australia.

House of Representatives Select Committee into the recent Australian bushfires, 2003. A Nation Charred: Report on the Inquiry of Bushfires. Commonwealth Australia, Canberra.

Keelty, M.J., 2011. A Shared Responsibility: The Report of the Perth Hills Bushfire February 2011 Review. Government of Western Australia, Perth.

Keelty, M.J., 2012. Appreciating the Risk: Report of the Special Inquiry into the November 2011 Margaret River Bushfire. Government of Western Australia, Perth.

Latz, P., 2004. Bushfires and Bush Tucker: Aboriginal Plant Use in Australia. IAD Press, Alice Springs.

Lewis, A.A., Cheney, P., Bell, D.E., 1994. A Report to the Hon Kevin J Minson by the Fire Review Panel Conducting a Review of the Department of Conservation and Land Management's (CALM) Prescribed Burning Policy and Practices and Wildfire Threat Analysis as Required in the Ministerial Conditions Set for the Implementation of Amendments to the 1987 Forest Management Plans and Timber Strategy and Proposals to Meet Environmental Conditions on the Regional Plans and the WACAP ERMP. Government of Western Australia, Perth.

Lindell, M.K., Arlikatti, S., Prater, C.S., 2009. Why do people do what they do to protect against earthquake risk: Perception of hazard adjustment attributes. Risk Analysis 29, 1072−1088.

Marsden-Smedley, J.B., 2009. Planned Burning in Tasmania: Operational Guidelines and Review of Current Knowledge. Tasmanian Fire Research Fund. Parks & Wildlife Service, Hobart.

McGee, T.K., Russell, S., 2003. "It's just a natural way of life…" an investigation of wildfire preparedness in Australia. Environmental Hazards 5, 1−12.

Mortimer, E., Bergin, A., Carter, R., 2011. Sharing risk: Financing Australia's disaster resilience. Australian Strategic Policy Institute Special Report 37, 1−24.

Paton, D., Bürgelt, P.T., Prior, T., 2008. Living with Bushfire risk: social and environmental influences on preparedness. Aust. J. Emerg. Manage. 23, 41−48.

Pyne, S., 2006. The Still Burning Bush. Scribe Publications, Carlton North.

Rodger, G.J., 1961. Report of the Royal Commission Bush Fires of December, 1960 and January, February and March 1961 in Western Australia. Government of Western Australia, Perth.

Smith, R., 2011a. Report on—Investigation of the House Losses in the Margaret River Bushfire 23 November 2011. Department of Fire and Emergency Services, Perth.

Smith, R., 2011b. Final Report on—Investigation of the House Losses in the Roleystone/Kelmscott Bushfire 6 February 2011. Fire and Emergency Services Authority, Perth.

Standards Australia, 2009. Australian Standard 3959-Construction of Buildings in Bushfire-prone Areas. SAI Global, Australia, Sydney.

Teague, B., McLeod, R., Pascoe, S., 2010. 2009 Victoria Bushfires Royal Commission: Final Report. Victorian Bushfires Royal Commission, Parliament of Victoria, Melbourne.

Tibbits, A., Whittaker, J., 2007. Stay and defend or leave early: policy problems and experiences during the 2003 Victorian bushfires. Environ. Hazards 7 (4), 283−290.

Ward, D., 2006. Resolving the Grasstree Fire History Dispute for the Benefit of Both Nature and Humans. Report to Hon. M McGowan as a Supplement to the Report "Dating Fires from Balga Stems: The Controversy Over Fire Histories Determined from South-western Australia *Xanthorrhoea*". Department of Environment & Conservation, Perth.

2009 Victorian Bushfires Royal Commission, 2010. Final Report. Government Printer for the State of Victoria, Melbourne.

Chapter 7

Fostering Community Participation to Wildfire: Experiences from Indonesia

Saut Sagala
School of Architecture, Planning, and Policy Development, ITB, Indonesia

Efraim Sitinjak and Dodon Yamin
Resilience Development Initiative, Bandung, Indonesia

ABSTRACT

Wildfire causes a lot of impacts to Indonesian forest areas. The impact includes loss of biodiversity, destroyed agriculture areas, and loss of properties. Communities living near forest areas are also at risk from wildfire impacts. This paper explores current practices of community preparedness to wildfire in Indonesia. The majority of wildfire occurrences in Indonesia in fact are induced by anthropogenic activities. Therefore, raising community preparedness will help not only to increase community resilience but also to their better management to mitigate fire occurrences. This paper argues that collaboration between stakeholders (community, private companies, and government) will be beneficial to increase not only community preparedness but also to reduce fire incidents.

7.1 INTRODUCTION

As a country located along the Equator and in the tropical region, Indonesia has the third largest tropical forest area in the world, after Brazil and Zaire (Makarim et al., 1998). These forest areas are mostly located in Sumatra, Kalimantan, and Papua Islands. In these forest areas, many indigenous people live near the forest. According to National Village Potential Data 2006 (Badan Pusat Statistik//National Statical Agency (BPS)) and Forest Area Map of 15 provinces, there are 1,305 villages (4.08 percent) of the total 31,957 villages in Indonesia that are in forest area 7,943 villages (24.86 percent) are located near forest area. These villages are mostly dependent to agriculture sector

Wildfire Hazards, Risks, and Disasters. http://dx.doi.org/10.1016/B978-0-12-410434-1.00007-5

(Forest Department, 2008). High levels of interaction between people and forest create two-way interactive relationships between forest and people which impact both forests and people. When a wildfire occurs near a village, it creates economic and health risks for its inhabitants. Indonesia's forests are a major component of the national economy, and provide significant wood-product exports, employment, domestic usage, and nontimber resources (Anderson et al., 1999). While forests continue to dominate the landscape in Indonesia, other land use types are expanding, including bush and scrub lands, grasslands, areas of cultivation, areas for permanent agriculture and settle-ments (Makarim et al., 1998). When an area of forest was opened or clear cut, some of the humidity or moisture from the soil would evaporate, and thus creating a drier condition of the soil which makes it more susceptible to fire, especially in peat soil. Moreover, the activity of clear-cutting or opening a forest would leave slash abandoned at the site, which is combustible and could cause potential harm of wildfires.

About 3.5 million hectares of Indonesian forests of 130 million hectares were heavily destroyed due to a long dry season in 1982 (Lennertz & Panzer,1984). This wildfire was also followed by fire incidents in 1983 which destroyed around 800,000 ha of primary forest; 1,400,000 ha of forest area that was cut for logging; 750,000 ha of secondary forest, agriculture field, and settlement; and 550,000 ha of wetlands and forest wetlands (Lennertz & Panzer,1984). Figure 7.1 shows the distribution of actual wildfire hazards in Indonesia (BNPB, 2009. Peta jumlah titik api (hot spot) di Indonesia. Jakarta).

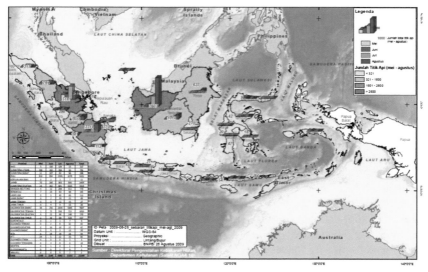

FIGURE 7.1   Wildfire hazards in Indonesia (BNPB, 2009).

According to data recorded between 1997 and 2013 (BNPB, 2013), areas susceptible to experiencing wildfire in Indonesia can be divided as follows: very high (Central Kalimantan, West Kalimantan, Riau, Riau Islands, and South Sumatra), high (East Kalimantan, Jambi, North Sumatra, and South Kalimantan), medium (Lampung, West Sumatra, East Jawa, Nangroe Aceh Darussalam, West Java, Central Java, Special Region Yogyakarta, West Nusa Tenggara, East Nusa Tenggara, and South Sulawesi), and low (Bengkulu, West Papua, Papua, Moluccas, DKI Jakarta, Central Sulawesi, North Sulawesi, South West Sulawesi, and Bali).

This chapter discusses two case studies, one in South Sumatra and the other in East Kalimantan, both of which are categorized in the "very high" and "high" category of wildfire susceptibility. Discussion of the characteristics of the communities in these provinces is used to explore how interdependency between community and forest influences risk and how it is managed. We start by reviewing wildfire causes and their management. Then, we discuss the community role in this process and how the history of wildfires in each area assists in understanding risk and its management. Finally, the role of government and private sector companies in wildfire management is discussed.

7.2 WILDFIRE AND WILDFIRE MANAGEMENT

Wildfires are a serious and growing hazard over much of the Indonesia, posing a considerable threat to life and property—particularly when they are located near built-up areas, such as settlements and agriculture fields. Secondary hazards, such as smoke, can cause health problems and environmental degradation. Yet, wildfires also constitute a natural process. Attempts at suppression in the past are now recognized to have created a larger fire hazard as living and dead vegetation accumulates in areas where fire has occurred.

According to General Directorate of PHPA (1994), there are three types of wildfires. The first type is ground fire. Ground fires usually happen in low land peat clay areas or in land where some minerals such as coal are found (Fredriksson, 2002). This type of wildfire is not easily detected. The second type is surface fires. Surface fires occur at the surface of the ground burning bushes, slash, and small trees. This type of fire does not burn the tree canopies since the trees are rare or resistant to fire. The third type is crown fire. Crown fires occur at tree canopies. The fire starts from the ground and proceeds to burn the tree branches and tree canopies. This type of fire is difficult to manage, especially when strong wind occurs. Peat soil in Sumatra covers considerably amount of area, covering some 30 percent of the total land area (Anwar et al., 1984), whereas in Kalimantan, the peat soil covers 10 percent of the total land area (Driessen et al., 1976). Thus, the occurrences of wildfire in both provinces are mainly peat fires which are hard to detect and to extinguish.

The majority of hotspots and wildfires have been caused by human activities usually associated with land use practices and changes

(Goldammer, 2011). Land use practices and changes consist mostly of clearing forests on a large scale to make space for producing key commodities such as palm oil and rubber on a large scale (Barber and Schweithelm, 2000).

While most of the wildfires in Indonesia are anthropogenic, glowing coal layers near the surface are also a potential wildfire source. Personal observations and interviews with villagers suggest that human-caused fires sweeping through the area often ignite such coal seams. Once lighted, these seams increase future fire risks by smoldering for a long time in an already degraded environment.

Looking at the major distribution of hotspots can help to predict wildfire occurrences. Hotspot detection can provide accurate and rapid information that increases awareness, alerts to mitigate, and assists managing wildfire (Saharjo, 2004). The number of hotspots varies every month according to the weather and climate (monthly rainfall pattern and maximum temperature), wind direction and speed, and types of vegetations that are burned. The drier the area, the higher the potential of hotspots occurring (Solichin, 2004). Therefore, normally wildfire occurs during the dry season.

Several research projects have studied the impact of vegetation fires in East Kalimantan and other Indonesian provinces (Goldammer et al., 2001; Hinrichs, 2000; Mayer, 1989; Schindele et al., 1989; Bappenas, 1999; Barber and Schweithelm, 2000; Schweithelm, 1999; State Ministry for Environment of the Republic of Indonesia and UNDP, 1998). Vegetation fires contribute to the degradation of fertile land quality and reduction of the socioeconomic value of nontimber resources from forest (e.g., rattan, honey, teakwood, clove, and agar wood). However, discussion of the role of participation of local community members in creating and managing wildfire risk is sparse.

7.2.1 Wildfire in Indonesia

As research on underlying causes of fire in Indonesia has indicated (Applegate et al., 2001), policy and institutional incentives (e.g., inappropriate land use allocation, lack of tenure security), and external forces (e.g., demographic changes) have influenced communities' use of their knowledge of fire behavior in sustaining their livelihoods. As depicted in Table 7.1, large areas of land and forest in Indonesia were burned in 1982 and 1983. Wildfires also burned during extended dry periods in 1987 (49,323 ha), 1991 (118,881 ha), and 1994 (161,798 ha). The fire areas in 1987, 1991, and 1994 were larger than during years with normal rainfall (Tacconi, 2003). The World Resources Institute (2013) noted that a large number of fire incidents are initiated at timber plantation and oil palm, accounting about 47 percent of the total incidents. A large number of timber and oil palm plantations indeed are found in Kalimantan and Sumatra, where the large number of fire hotspots occurred.

The historical data from 2001 to 2012 show that Sumatra Island suffers about 20,000 hotspot warnings every year (with the accuracy of 30 percent)

TABLE 7.1 Large Wildfire Incidents in Indonesia

Year	Location	Causes	Coverage Area	Losses
1982/1983	East Kalimantan	Long dry season	Destroying 3.2 juta ha	IDR 6 trillion
1987	West Region of Sumatra, Kalimantan until Timor East Area	Long dry season	In 1987, recorded 66,000 ha burn, in reality, it could have been more than that.	USD 9 million (estimate)
1991	West Region of Sumatra, Kalimantan until Timor East Area	Long dry season	500,000 ha with some reports of local smog (smoke & fog)	One Indonesian airline losses of 6.5 million and decreasing of hotels occupancy up to 70%. Forestry Losses N/A
1994/1995	Sumatra and Kalimantan	Long dry season	more than 5 million ha in 1994	Smog reaches up to Malaysia and Singapore
1997/1998	Almost in all area of Sumatra and Kalimantan	El nino and heat wave	1.3 million ha	USD 1.62–2.7 billion Some airports, sea ports, and road transportation were affected and closed. Pollution cost reached up to USD 674–799 million and related carbon emission of USD 2.8 billion.
2013	Riau, Kalimantan	Businessman, plantation	38,000–40,000 ha	IDR 10 trillion (estimate)

Source: FWI (2001), Bowen et al. (2001), Tacconi (2003), Boer (2002).

(WRI, 2013). The hotspot warning is normally issued between June and September every year. About 60 percent of the hotspots occur in the period of these months (WRI, 2013).[1] The large scale wildfires are induced by the climate change phenomenon, such as El-Nino that occurred in 1987, 1991, 1994, and 1997 (Environmental Ministry and UNDP, 1998).

In general, the impact of wildfire causes numerous health problems for the community. When the smog was very thick, the visibility of aircraft could be limited and thus hinder the aircraft from operating on time or even not operating at all. For the commercial fisheries, the limited visibility prevented the fishermen from going off shore. In normal condition, they would go on daily basis; however, with the existence of smog, they could only go once a week (Kompas, 2014a). To humans, smog could cause mild impact, such as eye irritation, to more serious impacts, such as aggravating existing health problem, particularly respiratory problems, such as asthma and many kinds of lung infections. Those who are the most vulnerable are children. Therefore, in some cases in Indonesia, schools were often closed whenever the smog was thick. However, until now, no fatalities have been reported. This might be due to the sites of wildfires being remote from human settlement.

In addition, the smog could be transported to neighboring countries such as Singapore and Malaysia by strong winds (The Economist, 2013a). For instance, a notable incident that happened recently occurred period between June 12 and 23, 2013 (WRI, 2013). At this time, wildfire warnings and incidents have increased in Singapore. The impact of these fires was dramatic. On Friday, June 21, the Pollution Standards Index (PSI), which is used to measure air pollution in Singapore, rose sharply to a record number of 400. This is much higher than the maximum acceptable PSI of 100 (The Economist, 2013b). This incident coincided with occurrences of 93 hotspots in Riau province, West Sumatra (Tempo, 2013a), and 54 hotspots in South Sumatra (Tempo, 2013b). In fact, the offenders most suspected in both areas were local people and private companies.

7.2.2 Wildfire Community Management

The term "community-based" entails much more than community members fighting fires. It is important to recognize that community involvement covers a wide spectrum of situations ranging from potentially forced participation in an activity (coercion) to free and willing participation in actions developed by the actors themselves (empowerment). While traditional or local knowledge is crucial on its own, it is insufficient to ensure sound, effective fire management. Institutional structures—both within and beyond the community—and the capacity to apply the knowledge are needed. While pertinent, timely, and appropriate knowledge about fires is useful, it will be of little use without

---

1. World Resources Institute.

community institutions organizing and directing the application of the knowledge.

To provide guidance and direction in wildfire control activities, the Minister of Forestry has established the regulation of the Minister of Forestry Number: *P. 12/Menhut-II/2009 On Wildfire Control*. In addition, the involvement of institutions and the public can be expected to occur and control all activities. Analysis of interests, influence, and role of the parties can map the position of institutions in wildfire control activities of both government and nongovernment, at both central and regional levels, as well as institutional follow-up programs that will be done to manage the institutional role.

For community based fire management to be effective, three fundamental components need to be understood (Makarabhirom et al., 2002): first, ecology and wildfire behavior, particularly wildfire regimes; second, the community, particularly its needs and the behavior of its members; and third, the relationships between fire and the community. Integrated wildfire management is a community-based fire management that offers a support program to address several issues including modules for institutional development, fire management training, equipment use, and maintenance (Abberger et al., 2002). The development of fire management crews—or volunteer village fire brigades—is a decisive step toward institutionalizing village fire management. The major task of such crews is to prevent and suppress wildfires in the village, and to promote safe burning practices in slash-and-burn agriculture in coordination and cooperation with the village and district authorities.

Paton (2005, 2006) and Paton and Bishop (1996) emphasize the need for risk communication being based on community involvement which encourages discussion of the issues hazards created for community groups (e.g., religious groups, social groups) in a way that empowers community members to identify the implications of the danger to their activities and facilitate their ability to deal with issues. When emergency agency community members talk about the dangers involved, the level of trust, satisfaction with communication, risk acceptance, willingness to take responsibility for their own safety, and collective commitment to face the consequences of danger will increase.

It is clear that participation is important but not sufficient to provide a context for evaluating the information. (Paton et al. 2005) demonstrated that whether or not people prepared was a function of how people interpreted their relationship with the hazardous aspects of their environment. Community members also need to direct their participatory endeavors in ways that facilitate their ability to identify what they need to know (Paton et al., 2006). However, hazard education programs rarely require active and sustained community participation as a component in programs intended to encourage participation (Paton et al., 2006).

Perception of environmental risk can be influenced by other's views. Similarly, mitigation decisions are also affected by how the large community sees risks (Earle, 2004; Jakes et al., 2003; Lion et al., 2002; Poortinga and

Pidgeon, 2004). Participation may trigger sharing new information from interaction with peers and colearning of new skills by involvement in substantial discussions, developing social capital through frequent contacts, and a sense of collective action to improve collective quality of life (Dalton et al., 2007; Earle, 2004).

7.2.2.1 Research Design

Community case studies were carried out in two Indonesian provinces where many wildfire events have occurred in the past: South Sumatra and East Kalimantan (Figure 7.2). Documents for each of the case studies were obtained from government officials, donor agencies, and previous research studies in South Sumatra and East Kalimantan. In addition to the documents as the main information sources, observations of the characteristic of the local community and interviews with several government staff were conducted. The data collected in South Sumatra are based on a study done previously by Rahmania (unpublished), while those collected in East Kalimantan are based on a study by Bunna (unpublished). In South Sumatra, the research focuses at communities nearby and around rainforest area, Muba District at the border between Jambi and South Sumatra Provinces, while in East Kalimantan, the research was conducted in Lesan River Conservation Area, Berau District (Bunna, unpublished).

Figure 7.3 depicts the community characteristic in South Sumatra and East Kalimantan. As can be seen, both communities have similar characteristics. In both communities, the percentage of people work as farmer is low (5 percent in East Kalimantan and 3 percent in South Sumatra). The level of education in both communities is relatively low: the biggest percentage of people graduated only from elementary school (53 percent in East Kalimantan and 31 percent in South Sumatra). However, higher level of education is observed in South Sumatra, having higher percentage of people graduate from high school (42 percent in South Sumatra in comparison to 34 percent in East Kalimantan) (Bunna, unpublished; Rahmania,

FIGURE 7.2   Study location.

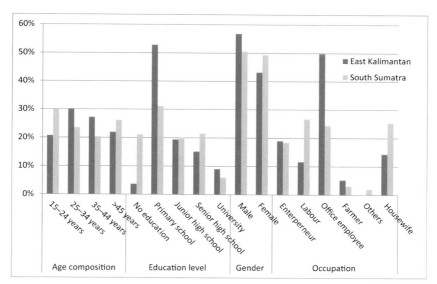

FIGURE 7.3 Community characteristics at South Sumatra and East Kalimantan (Rahmania, unpublished; Bunna, unpublished).

unpublished; BPS, 2010 (Indonesia in Figures 2010, Jakarta)). In addition, the percentage of poor people who live in rural area of East Kalimantan is 10.24 percent and of South Sumatra is 14 percent; with level of income of poor people in a month are 33USD and 23USD per person, respectively (Statical Agency/BPS, 2013. Indonesia in Figure, Jakarta). This implies that the community in East Kalimantan is slightly better off than those in South Sumatra.

This research used a mixed method research design employing qualitative and quantitative data collection methods. In the interviews, the main issues and problems experienced and lessons learned in each community were discussed. Interview data was complemented by observational data collected in each community.

Interviews were conducted with three main stakeholders: local community, government, and NGOs. Community data were collected from 89 interviewees in South Sumatra (Rahmania, unpublished) and 401 interviewees in East Kalimantan (Bunna, unpublished). Data were collected using a Knowledge, Attitudes, and Behavior survey. Demographic, psychographic, wildfire knowledge, and wildfire-related, attitude data were also collected.

In addition to the questionnaire data, we conducted a qualitative survey about wildfires hazards in Indonesia, South Sumatra, and East Kalimantan. The questions covered topics as follows: community activities that could trigger fires; community behavior in land burning; participation and anticipation of the community in preventing wildfires; and collaboration between the government, private companies, and community in reducing wildfires hazard. The NGOs interviewed were World Wildlife Fund and Indonesian Forum for Environment (known as Walhi). Although government official that were interviewed were those acting as field instructors at the Ministry of Forestry—Bangka Belitung and National Park at Bukit Barisan in Jambi, researchers at the Center for Research and Development on Forest Fires, Department of Forestry, South Sumatra, Forest Resource Production Sumsel.

7.3 COMMUNITY PARTICIPATION AND WILDFIRE IN SOUTH SUMATRA AND EAST KALIMANTAN

In Indonesia, wildfires are closely related to human activities. The main activities that ignite wildfires are those carried out by private companies, commonly by palm oil companies. Fewer wildfire ignitions can be attributed to the actions of local communities. Some of the incidents were purposely and consciously carried out to clear the land. Two areas that are most affected by wildfires are Sumatra and Kalimantan. In these islands, the soil is mainly peat, making the land prone to wildfires. Therefore, the participation of local community is urgently required to reduce the incidence of wildfires. In the next section, the wildfire hazard, as well as the participation of local community in South Sumatra and East Kalimantan wildfire risk management, is discussed, followed by a brief analysis of the role of government and private companies in wildfire risk management.

7.3.1 Wildfire in South Sumatra

South Sumatra Province covers a total area of 8.7 million hectares of which about 4.4 million hectares is covered by forests (Ministry of Forestry, 2009. Data and Information of South Sumatera Forestry, Jakarta). The forest composition consists of 17.22 percent of protected forests, 16.17 percent of conservation forest, 51.92 percent of permanent production forest, 9.77 percent of convertible production forest, and 4.92 percent is a limited production forest. Most of the forest area is steadily declining due to higher rates of deforestation. Average annual deforestation rate from 1985 to 1998 was approximately 192,824 ha/year (McCarty, 2000). High deforestation is driven by several community activities and policies issued by the government (Chokkalingam et al., 2001). One major cause of deforestation is high trans-migration programs, granting mining licenses, plantations, and community behavior (CIFOR, 2003).

According to Ramon and Wall (1998), the extent of wildfires in 1997/1998 covered about 34,229 ha of forest and about 19,318 ha of nonforest area. Figure 7.3 shows the increasing number of hotspots in South Sumatra in 2001, 2004, and 2012. Fire hotspots increased due to converting forests to palm oil plantations and forest production activities (Solichin, 2004; Wardani, unpublished). The occurrences of hotspots in the case study area, Banyuasin Districts, is categorized as "very high" in South Sumatra (Figure 7.4). Another area that has a similar characteristic is *ogan komering ilir* (OKI). The occurrence of wildfire in these districts is high because they have the largest peatland areas in South Sumatra.

Research suggests several reasons for the increase in wildfire risk in South Sumatra Province, including land clearing by people and large companies, specific soil conditions, and more extreme climatic conditions such as long droughts that occur annually from July to October (3—5 months each year) (Bowen et al., 2000). Burning forests by the public is generally carried out by low-income communities, especially on the east coast of South Sumatra (Fakhri, unpublished). They do not have much choice but to use fire as a means of a cheap, easy, and fast way for supporting life-sustaining activities (Septicorini, unpublished). They use the newly opened land for plantation field, rice field, and fish farming (Baharuddin, 2001).

However, a wider area was burned by private companies. This finding was pointed out by Bowen et al. (2000), who showed that another cause of wildfires in South Sumatra is associated with logging and forest businesses run by several large companies. These also include land use conversion by some oil palm and coal mining companies. Although the government already set some regulations prohibiting illegal logging and land burning, the absence of law enforcement in these areas has resulted in these activities continuing. Further, those caught red-handed received lenient punishment (i.e., retaining the permission to build a house; penalty), thus reducing the deterrent effect of this legislation (Kompas, 2014a). Peat wetland covers about 30 percent of the area of South Sumatra Province (Anwar et al., 1984). Because South Sumatra is positioned in tropical zones leading to an equal split between wet and dry seasons, peat lands in South Sumatra Province are susceptible to catastrophic wildfires.

7.3.1.1 Community Participation to Wildfire in South Sumatra

Forest in South Sumatra is an important community asset. Community perception of forest function can be categorized into forest as a source of livelihood (28 percent) and as a place to live in (28 percent). Forest is the source of livelihood because it fulfills a variety of community needs such as firewood, building materials for houses, and source of water that maintains the balance of nature (Figure 7.5).

The community is the key to the survival of forests through integrating indigenous knowledge, conservation values, and sustainable livelihoods (Moore et al, 2002). Managing forests using community involvement is more effective, especially if it is an entrenched social responsibility (Chamarik and Santasombut, 1994; Wasee, 1996; Sukwong, 1998; Ganz et al., 2001). An example of traditional knowledge of fire management is to burn the

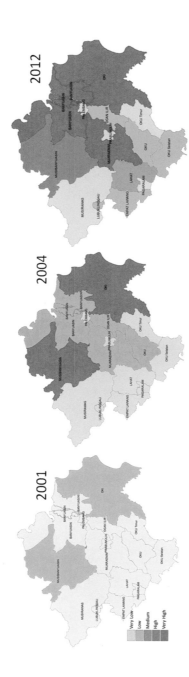

FIGURE 7.4   The increasing of hot spots in South Sumatra in 2001, 2004 and 2012 (Lapan, 2014. Site location, http://www.lapan.go.id (accessed in April 2014)).

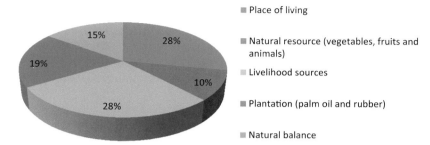

FIGURE 7.5   Opinion on forest function for the community (Rahmania, unpublished).

surrounding area encircle the fire source to stop approaching fires. However, the villagers tend to neglect the potential damage of land burning for economic reasons, as this quote shows:

Economic conditions and the knowledge of the society is the reason why people open the land by burning the forests. Land clearing costs by burning very inexpensive, does not require a lot of power, and its speed has become people's main reasons
          Staff, Center for Forest Natural Resource Conservation, South Sumatra, 2013

Similarly, Setijono (2001) in Rahmania (unpublished) confirmed that most people use fire for economic (29 percent), and practical (21 percent) reasons. Other reasons they identified include that clearing the forest through controlled burns is more effective and faster (36 percent) and that wildfires can make the soil more fertile (13 percent).

Wildfires can damage harvests, farmland, houses, and properties. Wildfires can also cause health problems, injuries, and deaths. Public perception of the importance of the role of the community to reduce the risk of catastrophic wildfires is crucial for community participation and engagement in fire prevention and elimination. Based on research conducted in the area OKI South Sumatra by Soewarso (unpublished), people who depend on forests and produce from plantations close to forests will be more motivated to engage in fire prevention and elimination. Therefore, in such a society, there is an unwritten rule related to compensation for any wildfire activity. Community perceptions also shape the behavior of people during the dry season. For example, in the dry season, people do not conduct common activities that use fire such as cooking and smoking (Soewarso, unpublished). They conduct preventive measures such as building canals around their gardens and keeping their gardens wet in the dry season to prevent any fire ignited (Giesen, 1991; Brady, 1997; Anderson et al., 1999).

The role institutions play in preparing for and responding to wildfires is small. Institutional formal and nonformal policies are needed to regulate community actions. According to Kasih (unpublished), institutions encompass family, cultural institution (e.g., extended family gathering, kinship, tribal-based organization), and external institutions (e.g., donors, Non Goverment Organization (NGO)). Village communities living in or close to forest areas generally have a passion for obtaining information specifically related to the use of forest resources that do not damage the environment. However, their limited economic resources make them resort to land burning to clear the area for agricultural purposes.

Communities generally have their own specific rules and procedures for wildfire risk reduction and conservation. One example is a specific compensation rule in the subdistrict of Tulung Selapan in the OKI District. Another is that community controlled burning should only be done on one's own land. If any member of a community trespasses into another person's property, then they will be given a penalty (Suyanto, 2002). In the subdistrict Pedamaran Kayuagung in the OKI district some areas are seen as sacred, and often, a myth is associated with these areas that ensures that these areas are protected. As a result, forests of this region tend to be maintained due to people's fear of being cursed, with this belief being sustained by community norms and practices.

7.3.2 Wildfires in East Kalimantan

From 1990 to 2010, in districts in East Kalimantan, such as Kutai Kartanegara, Kabupaten Kutai Timur, Balikpapan, Pasir, and Nunukan, intact forests were cleared to establish oil palm plantations. Currently, palm oil plantations cover an area of 31,640 square kilometers, which is an increase of 300 percent since 2000. In the same time frame, in Kalimantan, 47 percent of oil palm plantations were created on cleared forestland, 22 percent on secondary forest, and 21 percent on a mixture of forest and farmland (Carlson et al, 2012). Only 10 percent of palm oil plantations were established on nonforest area. These forests converted to palm oil plantation clearly reduce the biodiversity of vegetation and animals, and thus decrease the natural resources that are needed to support local communities (Abberger et al., 1999). Moreover, the common practice of palm oil companies clearing plantation area by burning the land is therefore increasing the potential risk of wildfire.

Natural resources are still the main source of people's livelihood along the River Lesan. The 12,192 ha of land along the Lesan River is protected conservation area (Bunna, unpublished). The villages near Lesan River Conservation area are inhabited by a wide variety of indigenous ethnic groups (Bina Organization, 2006). Communities in four villages depend on natural resources in the region for getting honey, *agar wood*, rattan, *resin*, genius leaves (palm), fish, and animals. Local people, especially from the Dayak Lesan Gaai, need access to timber and nontimber resources contained in the Lesan River Protected areas for the purposes of the village needs (subsistence) as for building materials, boat (*katinting*), consumption, and others. The decline of natural resources due to land use change makes it increasingly harder for people to generate livelihoods in this area

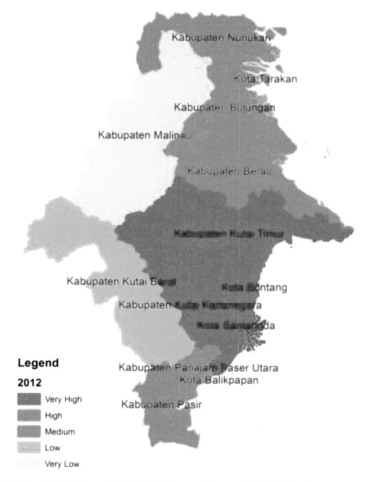

FIGURE 7.6   Wildlife hazards in East Kalimantan (Lapan, 2012. Site location, http://www.lapan.go.id (accessed in November 2013)).

(Bunna, unpublished). The land use change is the result of a regulation issued by the Governor of East Kalimantan No. 521/9038/EK on November 10, 2005[2] (Figure 7.6).

2. According to the recommendation from the Governor of East Kalimantan No. 521/9038/EK on November 10, 2005, about the Conversion of Land Use—directed to the Ministry of Forestry—the original land use that was converted into conservation forest was cultivation land used by local people. In this recommendation, a reduction of total area occurred for the forest conservation in Lesan river from 12.192 to 11.342 ha. A reduction of 850 ha was allocated to private companies for palm oil plantation by *PT. Belantara Pusaka*.

7.3.2.1 Community Participation and Fire risk management, East Kalimantan

In Kalimantan, indigenous communities have initiated wildfire management efforts in response to the perceived causes of many fires that adversely affected their community and agroecosystems during the late 1990s (Gemmingen, 1998). In general, most initiatives have focused on emergency wildfire suppression, rather than tackling the underlying causes of wildfires. The results of community based fire management strategies are contested; there are different perspectives on what constitute successful results (Abberger et al., 2002).

Another threat to forest preservation in Indonesia is illegal logging (Figure 7.7). The increasing illegal logging activity has increased the risk of wildfires and land use conversion, as with less trees, the land would lose its moisture and become drier, communities living in the Lesan River protected forest area of Lesan River identified wildfires as one of the causes of forest destruction (Bunna, unpublished).

Communities know their lives are highly dependent to the forest (Bunna, unpublished). Communities also know that the direct benefits they derive from forests will be maintained when conservation is preserved (Boer, 2002). While communities support the conservation of forest resources and the establishment of protected areas along the river Lesan, they do not know the benefits of protected areas for them. They believe that these areas being protected means that they cannot use the natural resources these areas provide for them. As a result, the people seemed to hesitate to support the program and tend to be indifferent. The wildfire management in the Lesan River Protected Area is managed by a collaborative approach involving government representatives at the district, subdistrict, and village.

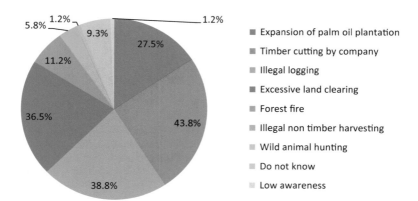

FIGURE 7.7   Reasons of destroyed forest in East Kalimantan (Bunna, unpublished).

7.4 COMMUNITY PARTICIPATION IN WILDFIRE MANAGEMENT

Taken together, our research suggests that private companies are at the center of wildfire management in Indonesia, because most wildfires are caused by, and related to human activities, which in this case are private companies. As depicted in Figure 7.8, community participation requires a concerted, interactive collaborative effort of communities, government, and the private sector. Collaboration eliminates limitations in term of resources, knowledge, and access (Cheng and Sturtevant, 2011). Therefore, local communities which are living in or close to forests could work side by side with the government in overseeing the activities and violation of law in the forest. This might be appropriate because communities have vital resources, including knowledge and understanding of the forest condition, in contrast to the government that is lacking manpower to cover vast area of forest, that contribute to sustainable forest use. Accordingly, the appropriate roles for government are law enforcement in relation to the activities of the private companies, and acting as facilitator and motivator for local community. For instance, government should facilitate development in low income communities by providing means to organize regular patrol and thus reduce the incidents of illegal logging and land burning.

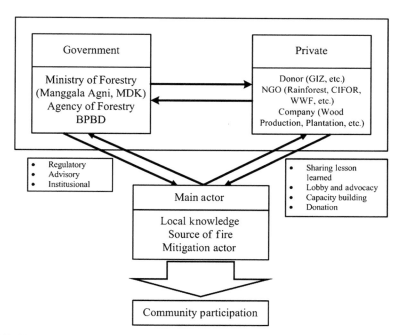

FIGURE 7.8 Interaction between community with other stakeholders that promotes participation.

This finding is in line with the community forest scheme (*Hutan Kema-syarakatan*) proposed by Indonesian Ministry of Forestry. The scheme is also known as *Hutan Taman Rakyat* (HTR) which literally translates as community garden forest (Kompas, 2014b). The scheme is arranged by law in Chapter 47 of Regulation by Ministry of Forestry, Number P.49/Menhut-II/2008 about *Hutan Desa* or District Forest. The community forest scheme is designed so that the local communities that live nearby or in the forest would be the ones who manage and take charge of the forest. Previously, the community was not entitled to have access to the forest. They could not enter into the forest to cut down or plant trees, since forest at that time was regarded as a property of the government or known as *Hutan Negara*. With this scheme, the community gain not only access to land but also rights to replant old trees with new ones and to grow plants that have economic value such as clove tree, rubber tree, agar wood, or teakwood. As regulated by law, the government even should provide some of the seeds of the new plants. Consequently, the community is required to commit to forest conservation for the regulated time span (≥35 years) (Kompas, 2014b).

However, only a small percentage of the forest areas have been converted into community forest, whereas most of the area still belong to the government. Moreover, it was reported that local communities have difficulties in applying for the scheme due to a lengthy bureaucracy, with the associated taking a year to take effect (Kompas, 2014b). Therefore, the government should simplify the process to ensure that the conservation program is fully implemented. However, there is increasing interest from the palm oil companies in using the peat areas, and thus creating a conflict of interest between conservation and commercial use. Therefore, the government should take charge in determining the border between land uses, monitoring, and the peat soil rehabilitation and conservation.

Apart from government support, the private sector, including plantation companies, also need to be involved to strengthen community participation. As noted, private companies account for many land use changes from forest to plantation areas, which increase their susceptibility to wildfires. Given that these companies greatly benefit economically from using the land, they have an interest in increasing the protection of their plantations, have a social responsibility to contribute to participation, and have the financial means to contribute. For example, National Resources Conservation Agency (known as *BKSDA*) of South Sumatra established *Mangala Agni*, which is a community-based unit of wildfire fighters.

REFERENCES

Anwar, J., Damanik, S.J., Hisyam, N., Whitten, A.J., 1984. Sumatra Ecosystem Ecology. Gadjah Mada Univ. Press, Jogyakarta, pp. 245–251 (in Bahasa).

Abberger, H., Sanders, B.M., Dotzauer, H., 2002. The development of a community-based approach for an integrated wildfire management system in East Kalimantan, Indonesia. In:

Proceedings of an International Conference on Community Involvement. Fire Management. FAO, Bangkok.

Anderson, I.P., Bowen, M.R., Imanda, I.D., Muhnandar, 1999. Vegetation Fires in Indonesia: The Fire History of the Sumatra Provinces 1996−1998 as a Predictor of Future Areas at Risk. FFPCP (Wildfire Prevention and Control Project) Report, Palembang.

Applegate, G., Chokkalingam, U., Suyanto, S., 2001. The Underlying Causes and Impact of Fires in Southeast Asia. Final Report. Center for International Forestry Research (CIFOR), Jakarta.

Baharuddin, 2001. Local community perspective of fire problem related to community life in swamp area/peat area (In Air Sugihan, South Sumatera). In: Semiloka Proceeding, Palembang, December 10−11, 2003. FFPCP.

Bappenas (National Development Planning Agency), 1999. Causes, Extent, Impact, and Costs of the 1997/98 Fires and Drought. Final Report. Asian Development Bank Technical Assistance Grant TA 2999-INO. Planning for Fire Prevention and Drought Management Project, Jakarta, Indonesia.

Barber, C., Schweithelm, J., 2000. Trial by Fire: Forest Fires and Forest Policy in Indonesia's Era of Crisis and Reform. World Resources Institute, Washington DC.

Bina, Swadaya. 2006. Report of Feasibility Study Development Ecotourism in Lesan River as Protect Area (unpublished) Jakarta: Bina Swadaya

BNPB, 2009. Map of Fire Hotspots in Indonesia. Jakarta.

BNPB, 2013. National Contingency Planning to Prevent Smog from Wildfire of Forestry and Land. Rencana Kontinjensi Nasional Menghadapi Ancaman Bencana Asap Akibat Kebakaran Hutan dan Lahan, Jakarta.

Bowen, M.R., Bompard, J.M., Anderson, I.P., Guizol, P., Gouyon, A., 2000. Anthropogenic fires in Indonesia: a view from Sumatera. In: Radojevic, M., Eaton, P. (Eds.), Wildfires and Regional Haze in South East Asia. Nova Science, New York.

Bowen, M.R., Bompard, J.M., Anderson, I.P., Guizol, P., Gouyon, A., 2001. Anthropogenic fires in Indonesia, a view from Sumatra. In: Radojevic, M., Eaton, P. (Eds.), EU & Ministry of Forestry.

Boer, C., 2002. Forest and fire suppression in East Kalimantan, Indonesia. In: Proceedings of an International Conference on Community Involvement. Fire Management. FAO, Bangkok.

BPS, 2010. Indonesia in Figures 2010. Jakarta.

Brady, M.A., 1997. Organic Matter Dynamics of Coastal Peat Deposits in Sumatra, Indonesia (A Doctoral thesis). University of British Columbia.

Bunna, A. Communication Media Design for Education Conservation Based on Community Preferences and Their Effects on Community Change Knowledge, Attitudes and Behaviour in Sungai Lesan Protected Area, Berau, East Kalimantan. Institut Pertanian Bogor. Bogor, Unpublished.

Chamarik, Santasombut (Eds.), 1994. Community forest in Thailand: development trend Copy No. 1. Local Development Institute, Bangkok.

Cheng, A., Sturtevant, V.E., 2011. A framework for assessing collaborative capacity in community-based public forest management. Environ. Manage. 49, 675−689 (2012) Springer.

Chokkalingam, U., et al., 2001. Fire management, natural resource change and its effect on community life at the swamp area/peat-Southern Sumatra. In: Semiloka Proceeding, Palembang, December 2003, pp. 10−11. Center for International Forestry Research (CIFOR).

CIFOR, 2003. The fire of swampland/peat in Sumatra: problems and solutions. In: Proceeding of Workshop, Center for International Forestry Research (CIFOR).

Dalton, D.R., Hitt, M.A., Certo, S.T., Dalton, C.M., 2007. 'The fundamental agency problem and its mitigation: independence, equity and the market for corporate control'. Academy of Management Annals 1, 1−64.

Driessen, P.M., Buurman, P., Permadhy, 1976. The influence of shifting cultivation on a 'podzolic' soil from Central Kalimantan. In: Peat and Podzolic Soils and their Potential for Agriculture in Indonesia. Soil Research Institute, Bogor, Indonesia, Bulletin No. 3, pp. 95−115.

Earle, T.C., 2004. Thinking aloud about trust: A protocol analysis of trust in risk management. Risk Analysis 24 (1), 169−183.

The Economist, 2013a. Fires in Sumatra Eternal Flames (accessed in 05.07.13).

The Economist, 2013b. Smog over Singapore Hazed and Confused (accessed in 05.07.13).

Fakhri, U. Wildfire Rehabilitation Priority Areas Case Study: South Sumatera. Universitas Indonesia, Depok, Unpublished.

Forest Watch Indonesia, 2001. Portrait of Indonesian Forest (Potret Keadaan Hutan Indonesia). Global Forest Watch, Bogor.

Forest Department, 2008. Site location. http://www.dephut.go.id/index.php/news/details/4914 (accessed in November 2013).

Fredriksson, G., 2002. Extinguishing the 1998 wildfires and subsequent coal fires in the Sungai Wain Protection Forest, East Kalimantan, Indonesia. In: Proceedings of an International Conference on Community Involvement. Fire Management. FAO, Bangkok.

Ganz, D.J., Moore, P., Shields, B., 2001. Workshop report. In: International Workshop on Community Based Fire Management. December 6−8, 2000, RECOFTC Training and Workshop Report Series, 2001/3. Bangkok.

Gemmingen, G., 1998. Rehabilitation of Fire-Affected Forests in East Kalimantan. Center for International Forestry Research (CIFOR) (IFFN No. 19-September 1998, pp. 23−26).

Giesen, W., Sukotjo, 1991. Conservation and Management of the Ogan-Komering and Lebaks South Sumatra. PHPA/AWB Sumatra Wetland Project Report No.8 Center for International Forestry Research (CIFOR).

Goldammer, J., Frost, P.G.H., Jurvélius, M., Kamminga, E.M., Kruger, T., Moody, S.I., Pogeyed, M., 2001. Community participation in integrated wildfire management: experiences from Africa, Asia and Europe. In: Proceedings of an International Conference on Community Involvement. Fire Management. FAO, Bangkok.

Goldammer, J.G., 2011. Wildland fires and human security: Challenges for fire management in the 21st century. Proceedings of the International Forest Fire Symposium Commemorating the International Year of Forests 2011, 7−8 June 2011 Sol Beach, Gangwon-do, Republic of Korea, 2011.(pp. 36−49): Korea Forest Reasearch Institute.

Hinrichs, A. 2000. Financial losses due to the 1997/8 fires in timber concession of East Kalimantan, Indonesia. Samarinda, Indonesia,GTZ-SFMP Report (unpublished).

Jakes, P.J., Nelson, K., Lang, E., Monroe, M., Agrawal, S., Kruger, L., Sturtevant, V., 2003. A model for improving community preparedness for wildfire. In: Jakes Compiler, P.J. (Ed.), Homeowners, Communities, and Wildfire: Science Findings from the National Fire Plan. Proceedings from the 9th International Symposium on Society and Resource Management, June 2002, Bloomington, IN, pp. 4−9.

Kasih, N. The Concept of Community Empowerment Kerinci Seblat National Park Zone (TNKS) Musi Rawas District, South Sumatera. Bogor Agricultural Institute, Bogor, Unpublished.

Kompas, 2014a. New Hope from Barru Forest (accessed 24.02.14.).

Kompas, 2014b. Chaos in Flight Schedule in Pekanbaru (Last accessed in 24.02.14).

Lennertz, R., Panzer, K.F., 1984. Preliminary assessment of the drought and forest fire damage in Kalimantan Timur. Report by DFS German Forest Inventory Service Ltd for Deutsche Gesellschaft für Technische Zusammenarbeit (GTZ).

Lion, R., Meertens, R.M., Bot, I., 2002. Priorities in information desire about unknown risks. Risk Analysis 22, 765−776.

Makarim, N., et al., 1998. Assessment of 1997 Land and Forest Fires in Indonesia: National Coordination. IFFN No. 18-January 1998, pp. 4–12. Jakarta.

Makarabhirom, P., Ganz, D., Onprom, S., 2002. Community involvement in Fire Management: cases and recomendations for community-based fire management in Thailand. In: Moore, P., Ganz, D., Tan, L.C., Enters, T., Durst, P. (Eds.).

Mayer, J.H., 1989. Socioeconomic aspects of the forest fire 1982/83 and the relation of local communities towards forestry and forest management in East Kalimantan. FR-Report No. 9. German Forest Service (DFS), Feldkirchen, Germany.

McCarthy, J.F., 2000. "Wild Logging": The Rise and Fall of Logging Networks and Biodiversity Conservation Projects on Sumatra's Rainforest Frontier. CIFOR Occasional Paper No. 31. Center for International Forestry Research, Bogor.

Ministry of Forestry, 2008. Site location. http://www.dephut.go.id/index.php/news/details/4914 (accessed in November 2013).

Ministry of Forest. 2009, Data and Information of South Sumatera Forestry, Jakarta.

Moore, P., David, G., Tan, L.C., Enters, T., Durst, P.B., et al., 2002. Communities in flames. In: Proceedings of an International Conference on Community Involvement in Fire Management. FAO, Bangkok.

Paton, D., Bishop, B., 1996. Disasters and communities: Promoting psychosocial well-being. In: Paton, D., Long, N. (Eds.), Psychological Aspects of Disaster: Impact, Coping, and Intervention. Dunmore Press, Palmerston North.

Paton, D., Smith, L.M., Johnston, D., 2005. When good intentions turn bad: Promoting natural hazard preparedness. The Australian Journal of Emergency Management 20 (1), 25–30. ISSN 1324-1540.

Paton, D., 2005. Positive adaptation to disaster and traumatic consequences: Resilience and readiness. The Australasian Journal of Disaster and Trauma Studies 2, 1–3.

Paton, D, Disaster response and recovery: Considering volunteers, displaced communities and cultural heritage, The Australasian Journal of Disaster and Trauma Studies, 2006-1 pp. 1–2. ISSN 1174-4707.

Paton, D., Gail, K., Buergelt, P., Michael, D., 2006. Preparing for bushfires: understanding intentions. Disaster Prev. Manage. 15 (4), 566–575.

Poortinga, W., Pidgeon, N.F., 2004. Trust, the asymmetry principle, and the role of prior beliefs. Risk Analysis 24, 1475–1486.

Rahmania, 2011. Community Perception of the Ecosystem Restoration Activities in the Area of Harapan Rainforest PT REKI, Jambi and South Sumatra Province. Bogor Agricultural Institute, Bogor, Unpublished.

Ramon, J., Wall, D., Forest Fire Prevention and Control Project, European Commission, 1998. Fire and Smoke Occurrence in Relation to Vegetation and Land Use in South Sumatra Province, Indonesia with Particular Reference to 1997. Report No. 47. Palembang: European Commission's Wildfire Prevention and Control Project.

Saharjo, B.H., 2004. Adaptation of Forest Development Planning under the Shadow of Climate Change (Adaptasi Rencana Pembangunan Kehutanan Dibawah Bayang-bayang Perubahan Iklim) (Forest Fire and Land Laboratory and Faculty of Forestry, Bogor Agriculture University (Laboratorium Kebakaran Hutan dan Lahan. Fakultas Kehutanan Institut Pertanian Bogor).

Schindele, W., Thoma, W., Panzer, K., 1989. The forest fire 1982/83 in East Kalimantan. Part 1: The fire, the effects, the damage and technical solutions. German Forest Service (DFS), Feldkirchen, Germany.

Schweithelm, J., 1999. The fire this time: an overview of Indonesia's forest fire in 1997/1998. WWF Indonesia Forest Fire Project, WWF Indonesia, Jakarta, Indonesia.

Septicorini, E. The Determination Level of Vulnerability Study of the Wildfire in Ogan Ilir Komering of South Sumatra Province. Bogor Agricultural Institute, Bogor, Unpublished.

Setijono, D., 2001. Kehidupan community and relation to fire swampland/peat in OganKomering ilir of South Sumatra province. In: Proceedings of the Workshop, pp. 10–11. Palembang December 2003.

Soewarso. Preparation of Peat Swamp Wildfire Prevention Using Prediction Model (Case Study: Forest Group of Sungai Sugihan and Sungai Lumpur, South Sumatra Province). Bogor Agricultural Institute, Bogor, Unpublished.

Solichin, 2004. The Tendency of the Wildfire in South Sumatra: Analysis of Historical Data NOAA and MODIS Hotspot. South Sumatera Wildfire Management Project. Palembang.

State Ministry for Environment of the Republic of Indonesia and UNDP, 1998. Forest and land fires in Indonesia. Volume I: Impacts, factors and evaluation. Jakarta, Indonesia.

State Ministry for Environment of the Republic of Indonesia & UNDP, 1998. Forest and land fires in Indonesia. Volume II: Plan of action for fire disaster management. Jakarta, Indonesia.

Statical Agency/BPS, 2013. Indonesia in Figure. Jakarta.

Sukwong, Somsak, 1998. Local culture "Khao Mor Kang Mor" for fighting forest fire. In Community Forest Newsletter 5 (10), 13–15. RECOFTC, Bangkok.

Sukwong, Somsak, 1998. Local culture "Khao Mor Kang Mor" for fighting forest fire. In: Community Forest Newsletter, 5. RECOFTC, Bangkok, pp. 13–15, 10.

Suyanto, S., Permana, R.P., Khususiyah, N., 2002. Fire, Livelihood and Swamp Management: Evidence from Southern Sumatra, Indonesia. Project Report. CIFOR, ICRAF and European Union, Bogor Indonesia.

Tacconi, L., 2003. Fires in Indonesia: Causes, Costs and Policy Implications. CIFOR Occasional Paper No. 38. Bogor.

Tempo, 2013a. Forestry Berau Monitor 19 Companies Smog Suspected (accessed in 24.06.13).

Tempo, 2013b. Police Catch Villager Who Burned the Land in Riau (accessed in 24.06.13).

Wardani, S. Studies Monthly Heat Distribution Point (Hotspot) as the Probe Wildfires in South Sumatra Province in 2001 and 2002. Bogor Agricultural Institute, Bogor, Unpublished.

Wasee, 1996. Community forestry development in Thailand. RECOFTC, Bangkok, pp. 27–34.

World Resources Institute, 2013. Site location. http://www.wri.org/ (accessed in November 2013).

Chapter 8

Discourse on Taiwanese Forest Fires

Jan-Chang Chen
Assistant Professor, Department of Forestry, National Pingtung University of Science and Technology, Pingtung, Taiwan

Chaur-Tzuhn Chen
Professor, Department of Forestry, National Pingtung University of Science and Technology, Pingtung, Taiwan

ABSTRACT

This chapter discusses how forest fires are a relatively new hazard in Taiwan, and one that is viewed by Taiwanese forestry agencies as a major hazard. This chapter explores how the need to rapidly understand fires that occur in remote, mountainous regions has focused attention on research and practices based on mapping that enable postdisaster, environmental monitoring and management, and how the rise in conservation awareness in recent years has focused attention on post-forest-fire restoration work and issues that have arisen regarding the implementation of traditional forest restoration after fires. This chapter also discusses the role of fire danger ratings and the benefits that could accrue from more international standardization in danger-rating systems.

8.1 INTRODUCTION

Forest fires present a significant challenge to forest protection in Taiwan, as elsewhere. While their effects are diverse, and can result in ecological benefits and social losses, forest protection requires that the hazard be managed. If a forest fire cannot be promptly extinguished, fire can create inestimable ecological, economic, and social losses and requires long periods of time, as well as capital investment and labor, to restore forests to their original form (Yen, 2004). Consequently, forest fires are viewed by Taiwanese forestry agencies as major disasters.

The complexities of forest fire management have been fueled by experience. For example, in Taiwan, the 2001 Lishan forest fire affected approximately 260 ha of forest land. After the fire, the Forestry Bureau planned to restore the

Wildfire Hazards, Risks, and Disasters. http://dx.doi.org/10.1016/B978-0-12-410434-1.00008-7
145

forest. However, the need to include social and amenity issues (e.g., consideration of views along the Central Cross-Island Highway) caused doubts regarding the straightforward implementation of traditional forest restoration practices after fires. The 2002 Wuling forest fire affected >300 ha of forest land. Not only did it destroy a forest that had been restored for 30 years but it also impacted a Formosan landlocked salmon protection area. After the fire, forest restoration plans once again caused debate (Lin et al., 2005).

The rise in conservation awareness has focused attention on post-forest-fire restoration work. This is facilitated by, for example, developing maps to enable postdisaster environmental monitoring and management (Hsieh et al., 2011) and by developing rating systems.

8.1.1 Forest Fire Ratings

Forest fires are often rated according to the size of the area affected or the amount of damage (in monetary terms). However, the rating systems used domestically and internationally are inconsistent. The rating systems used in the United States, China, and Taiwan are organized and shown in Table 8.1 (Chang, 1999; Fuller, 1991).

Table 8.1 illustrates how Taiwan's Forestry Bureau defines a fire alarm as a fire that damages an area of ≤ 5 ha, and causes monetary losses of below NT\$ 200,000. Fires defined as fire alarms in Taiwan are five times the area of fire alarms in China, and cover Classes A, B, and C in the United States. In Taiwan, the definition of a forest fire is a fire that affects areas >5 ha. This is considerably larger than China's definition of a forest fire (≥ 1 ha), and is equivalent to a Class C fire in the United States.

According to the forest fire statistics, Hwang, and Lin (2011), forest fires in Taiwan mostly occurred between January and April, followed by July, August, November, and December. Forest fires are least likely to occur in May, June, September, and October.

TABLE 8.1 Forest Fire Ratings

Country	United States	China	Taiwan
Rating	Class A: 0.02–0.10 ha Class B: 0.11–0.04 ha Class C: 4.05–40.43 ha Class D: 40.44–121.39 ha Class E: 121.40 ha or above	Fire alarm: 1 ha or below Forest fire: 1–100 ha Major fire: 101–1,000 ha Extremely large forest fire: $\geq 1,000$ ha	Fire alarm: 5 ha or below and damages less than NT\$ 200,000 Forest fire: 5 ha or above or damages exceeding NT\$ 200,000

8.1.2 Forest Fires in Taiwan

Statistical data on forest fires that occurred in Taiwan between 1992 and 2000 revealed 406 fire alarms and 243 fires. The damaged areas reached 5168 ha, and monetary damages were estimated at NT$ 825,045,000 (Forestry Bureau of the Council of Agriculture in the Executive Yuan, 2000). The Yushan Tataka fire in 1993, the Lishan fire in February 2002, the Yushan Dongfeng fire in March of the same year, and the Wuling fire in May 2003 were all large forest fires reported by the media, which made a deep impression on people. In Taiwan, 99.6 percent of forest fires are caused by human factors. The fact that 49 percent of arsonists who perpetrate fires are never discovered is regrettable. Human-caused fires can be attributed to agricultural burning, discarding lit cigarette butts, building fires for warmth and light, and burning of joss paper (Lin, 1994).

8.1.3 Forest Fire Prediction Method—Forest Fire Danger-Rating and Warning (Testing) System

Many factors influence forest fires, contribute differently to forest fire behavior, and highlight the importance of predicting forest fires. In 1930, researchers in the United States began developing a forest fire prediction system. Later, Canada and Australia joined the research. The forest danger-rating and prediction system currently in use is a revised version of the system that has been developed by these countries (Lin, 2001).

8.1.3.1 The Theory of Forest Fire Danger Rating

Two types of model frameworks are used to construct forest fire danger rating prediction models: static models and dynamic models (Wang, 2002). Static models divide areas into several forest fire danger zones based on fuel type and long-term weather data. Forests are located at various latitudes, and exhibit different terrain and climactic conditions. In addition, human socioeconomic factors also influence forest fire risk. Therefore, areas are accorded different forest fire danger ratings. Static models for forest fire danger rating prediction rate forest fire hazard areas on the basis of climate, terrain, vegetation type, and frequency of ignition source. These forest fire hazard zones can be used as a basis for establishing long-term or permanent fire prevention measures, and generally do not take short-term weather changes into consideration. Yang et al. (1992) divided the forest area in northeast China into units of forest compartments (based on forest resource survey data of forest resources, such as land type, slope, aspect, elevation, forest type, forest age, and density) to obtain potential forest fire hazard values for various forestry compartments. The forest fire danger level in the jurisdiction of each forestry compartment was then rated based on size. Lin (1994) divided national forestry areas in Taiwan based on forest vegetation composition, climactic zones, recreational systems, and forest fire frequency data. One of the analysis of geographical information system (GIS) was used for conducting spatial analysis to determine forest fire hazard areas in national forests. Potential forest fire hazard areas in national forests are divided into five classes.

Dynamic models take short-term weather and climactic factor changes, and the influence these have on the flammability of plants and other combustible items, into consideration. In these models, forest fire hazard factors are fixed, and the forest and surrounding plains are possibly flammable; their flammability can be influenced by atmospheric humidity, airflow intensity, and other factors at any given time. The flammability of fuel fluctuates drastically, and can change daily or hourly. Therefore, in this model, different weather factors and daily weather conditions are used to predict the ignition probability of fuel on a certain day or hour. This has considerable implications regarding temporary preventive measures or measures undertaken by fire fighters engaged in fighting the forest fire. Chang (2004) integrated a large quantity of experiment data based on water content of light-weight fuel and the initial speed of combustion with forest fire history and weather data, to conduct a statistical analysis and determine regularities. They then built a mathematical model and prediction method for forest fire danger rating. Currently, various countries worldwide employ forest fire warning systems suitable to their own natural backgrounds. The forestry departments of various countries regularly issue forest fire danger-rating notifications during fire season (Lin, 1992).

8.1.3.2 Forest Fire, Danger Rating, Warning System in Taiwan

Lin (1995) examined the feasibility of using Sweden's Angstrom index to test forest fire danger ratings and provided a reference for rebuilding Taiwan's forest fire danger rating model. Lin's research selected the daily recorded temperature and humidity from Taipingshan (dumping site) in the Luodong forest area, Mingchih, and Tiger monument, and analyzed and compared forest fires between 1982 and 1991. The results of the study showed that forest fires in Taiwan have a significant relationship with low temperature and humidity. The Swedish forest fire danger index calculation method is difficult to use for providing daily danger ratings. However, the accumulated information for five to seven days can still be used as a reference for short-term, fire danger prediction. The Forestry Bureau's forest fire danger-rating prediction system classifies danger rating into five levels according to the probability of fire in various areas.

8.1.4 Application of GIS and Aerial Telemetry Information in Forest Fire Predictions

Hsiao (2003) used the ignition mechanism in forest fires as the basis for estimating the time and space distribution of Taiwanese forest fire danger ratings. Hsiao et al. used forest fires and weather factors for forest fires that occurred in Taiwan between 1990 and 2001 as the basic data points. The current day's highest air temperature, temperature difference, accumulated time without rainfall, and drought index were used as independent variables, and forest fire events were used as dependent variables. A logistic regression model was used to build a forest fire ignition probability model. This model further considered the space and time variation in weather factors, and used GIS to conduct temperature and rainfall projections in space. The forest fire ignition probability model was spaced out in 1-km^2 network layers as a basis for projecting fire danger-rating predictions for forests in Taiwan on a given day. This model used data from between January and June 2002 to conduct tests. The results indicated that the accuracy of this model in predicting the occurrence of forest fires was 82.76 percent. In reality, the results of this study have already been practically used in the forest fire danger rating warning system built by the Forestry Bureau, which has functioned normally.

Chung et al. (2006) integrated GIS and moderate resolution imaging spectroradiometer (MODIS) imaging, and applied them to forest fire detection and to forest fire danger ratings. MODIS imaging was used to calculate the normalized difference vegetation index (NDVI), normalized thermal index, and normalized difference water index and build a forest fire detection model.

8.1.4.1 Forest Fire Danger Rating Index Model

GIS technology is used to render different layers of forest areas into the rating system. The spatial-analysis function of GIS is used to render the forest fire risk zone map, and to display danger ratings as a simple numerical value index. This is one of the methods that most closely match real-life situations. Several factors can be incorporated into the method for analysis that most commonly influence forest fires, including terrain, fuel type, human activities, and atmospheric conditions. With the technical advancement of information, the application of GIS to forest or disaster danger rating mapping has gradually improved (Chuvieco and Congalton, 1989). The framework for forest fire danger rating index can be summarized as the forest fire danger rating index framework in Figure 8.1 (Anderson, 1982).

1. Weather Danger Index: This is calculated based on value observed three times a day from a series of weather stations located in the forest test area. This calculation includes two factors: the probability of ignition (PI) and wind speed. The PI of dry fuel is calculated based on the atmospheric temperature and relative humidity, and the fuel exposure and distribution in the terrain. Wind speed in correlation with the dryness of fuel can be divided into three PI levels: medium-low, high, and extremely high.

2. Fuel Hazard Component: This refers to different vegetation growth patterns in relation to the expansion rate of forest fires. In a forest fire behavior system, this is divided into four categories (Anderson, 1982): forest fires on grassy plains, forest fires in brush or the humus layer of brush, forest fires in tree branches or brush, and forest fires in trees.

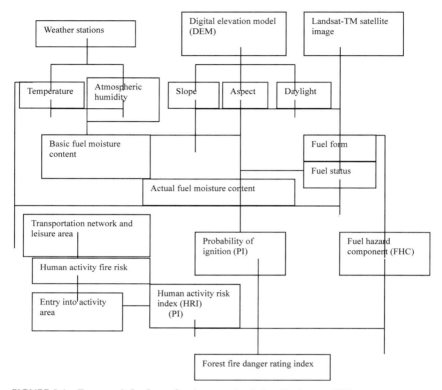

FIGURE 8.1   Framework for forest fire danger-rating index (Anderson, 1982).

3. Fire Incident Index (FII): This is obtained by calculating historical fire occurrence records. Another consideration is the human-activity risk index. This index represents the influence of human activities on these areas, such as unintentional fires caused by camping or agricultural activities, or intentional arson, which presents high forest fire risk (Chuvieco and Salas 1996).

8.1.4.2 Factors Included in the GIS Forest Fire Danger Rating Model

Based on the above-discussed forest fire danger index system, the following factors can be indicated: atmospheric temperature and relative humidity, slope, aspect, fuel form, fuel exposure pattern, and human risk factors. These factors can be digitized and mapped, and formed into a forest fire danger rating system. The GIS spatial-analysis function can be used to indicate an area's forest fire danger rating.

High-precision Digital elevation model (DEM) data was used to render topographic maps and produce slope and aspect strata (Figure 8.2). Although

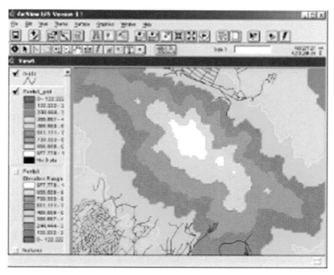

FIGURE 8.2   Elevation rating map rendered with GIS software.

slope and aspect do not directly influence the ignition of forest fires, in the Northern Hemisphere, the southern aspect is influenced by sun exposure, which dries fuel and renders it more susceptible to causing forest fires (Figure 8.3). Slope and wind influence forest fire behavior. Steep slopes and strong winds prompt fires to spread (Vakslis et al., 2004), whereas gentle slopes and mild breeze slow down fire spread (Figures 8.4 and 8.5). Weather data included temperature, relative humidity, and rainfall distribution, and were obtained from records of weather stations located inside the test area forest. GIS software was used to separately render the temperature gradient and rainfall-gradient layers.

Vegetation strata and vegetation patterns were classified by using satellite images or determined by aerial photography in coordination with ground survey data. This process allowed types of plant growth to be classified, fuel types in a given area to be determined, and fuel-pattern maps to be rendered (Figure 8.6). Further, the type of land use can be integrated into the vegetation strata and become one of the influencing factors.

Human-activity factors are the most complex. Generally, roads, railway tracks, leisure areas, and settlements are used to determine human-activity factors. The GIS buffer-analysis function is used to render different regional layers, based on the principle that human errors accidentally cause forest fires, and the distance involved (Figure 8.7).

8.1.4.3 Mapping of Forest Fire Danger Rating

GIS maintenance of data and common frameworks are helpful in displaying storing, and analyzing spatial data. However, GIS is unable to obtain collected

FIGURE 8.3   Aspect diagram (southern aspect slopes are more prone to forest fires, and have different vegetation, than northern aspect slopes).

FIGURE 8.4   Effects of wind and slope on forest fires; forest fires expand rapidly uphill.

data independently. Therefore, GIS must use the value maps in satellite images for rendering, and to input geographical data into the forest fire danger rating system (Agee and Pickford, 1985).

Generally, research on the mapping of fire hazard areas has used vegetation patterns, elevation, slope, aspect, and range of human activities (e.g., distance

FIGURE 8.5   Effects of wind and slope effects on forest fires; forest fires rapidly expand uphill (the yellow section is not influenced by these factors and shows a slow and even expansion).

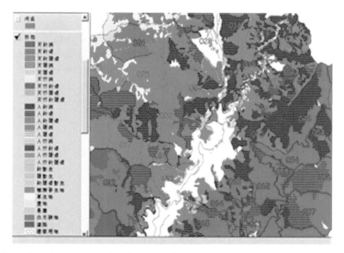

FIGURE 8.6   Forestry map, used to determine fuel type.

of settlements, roads, and camp sites as factors) as the basis for constructing layers, and have used the following procedure to build forest fire risk zone maps (Table 8.2):

1. Layers, which each have a theme, are divided by weights. Weights are designated based on corresponding forest fire risks. The final forest fire risk zone map is divided into four (or five) classes—very high, high, medium,

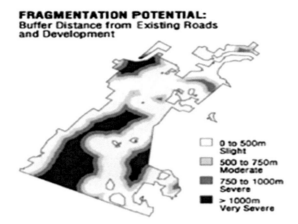

FRAGMENTATION POTENTIAL:
Buffer Distance from Existing Roads
and Development

0 to 500m
Slight

500 to 750m
Moderate

750 to 1000m
Severe

> 1000m
Very Severe

FIGURE 8.7   Roads and settlement buffer analysis (Richard and van Lear, 1998).

and low (or no risk, usually bodies of water) danger ratings—based on values obtained from weight selection and calculation results.

2. The data inside each theme layer are further subdivided into different levels; the coefficients of 0, 1, and 2 represent low, medium, and high levels of danger. This is used to calculate the following forest fire danger rating index, used as a basis to calculate forest fire danger ratings (H), and to determine the forest fire danger rating of each area: using the following formula:

$$H = 100V + 30s + 10a + 5r + 2e,$$

where v is the vegetation strata;
   s, the slope;
   a, the aspect;
   r, the distance of roads and settlements; and
   e is the elevation.

3. Based on the previously mentioned basic geographical information layer data, GIS software and techniques were used to conduct cross overlay analysis on different danger-rating primer layers, and produce the derived layer data. Finally, the forest fire, risk zone map is rendered (Figure 8.8). The spatial data layers (e.g., forest fire frequency and vegetation strata/fuel map) are derived from the forest fire history, ground data on fuel, and satellite-derived vegetation map in the basic GIS data layers. In addition, the digital terrain model, road network, and land use and land cover were used to estimate slope and the ignition source map. Finally, crossoverlay analysis was conducted on these basic layers and derived layers to produce the forest fire risk zone map.

TABLE 8.2 Classification and Weight Rating of the Forest Fire Danger Rating Model

Classification (Original Classes)	Fire Danger Rating (Fire Hazard Groups)	Coefficient (Coefficient)
Vegetation Strata Layers (Weight = 100)		
Dense coniferous forest	High	2
Medium density coniferous forest	High	2
Sparse coniferous forest and brush	Medium	1
Dense brush	Medium	1
Medium density brush	Medium	1
Sparse brush	Low	0
Apricot, almond (broad leaf) forest	Low	0
Vineyard	Low	0
Orange (fruit) orchard	Low	0
Slope Layers (Weight = 30)		
0–12%	Low	0
12–40%	Medium	1
>40%	High	2
Aspect Layers (Weight = 10)		
90–180°	High	2
180–270°	Medium	1
270–90°	Low	0
Distance to Neighboring Road Layer (Weight = 5)		
0–50 m	High	1
>50 m	Low	0
Elevation Layer (Weight = 2)		
0–400 m	Low	0
>400 m	High	1

TABLE 8.3 Forest Fire Ignition Probability

Ignition Probability	Danger Rating	Rating Description
0–20	1	Safe
21–40	2	Caution
41–60	3	Warning
61–80	4	Danger
81–100	5	Extreme danger

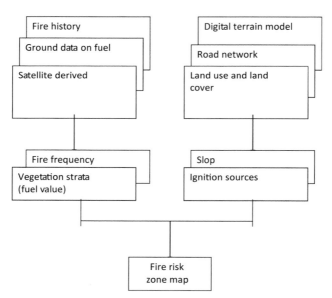

FIGURE 8.8   Procedure for producing forest fire risk zone maps (Abhineet et al., 1996).

The vegetation type, habitats, forest conditions, and vegetation strata of the several basic geographical information data layers previously mentioned in this study were determined by using supervised and unsupervised classification in Landsat TM. Elevation data were obtained through the digital terrain model interpolation method to calculate the elevation value of various points in the study area; these data were relatively inaccurate. In addition, slope and aspect-layer data were obtained by calculating the elevation data. Layer data for neighboring roads and settlements were digitized by GIS and stored in separate files. Buffer analysis in GIS was used to provide distance value settings and produce new area layers. All the layers were rasterized, and each grid was given a value based on weight. Finally, these different layers were overlaid with each other, and the corresponding value of each grid was totalled to render a forest fire risk zone map.

8.1.5 Using GIS Technology to Develop Decision-Support Systems

Generally speaking, forest fires severely damage the natural environment, and are of significant importance to neighboring natural environments and biological groups. Forest fire management and decision making are built on rational data collected from the ecological environment and data related to possible ramifications of fires. To compile such data, the development of an operating system that can manage large-scale, forest fire events is required; such an operating system must incorporate comprehensive GIS and relational database management system technology, and provide mutual interactions and contact abilities (Vakalis et al., 2004).

8.1.6 Applying GIS to Forest Fire Management—Using Current Forestry Bureau Methods as an Example

Taiwan has an island climate characterized by high humidity, and forest fires do not easily occur. However, because of various factors, statistical data shows that 30—40 forest fires occur annually. One major contributing factor is unintentional or intentional ignition caused by human activities. Taiwan has steep terrain; once a forest fire occurs, terrain limitations prevent fire fighting activities, causing large losses. Therefore, forest fire prevention, determining the fastest method for putting out forest fires, and minimizing losses are the objectives expected to be achieved by competent forestry authorities (Forestry Bureau).

Because of advances in computer technology, computers can be used to integrate geographical data and render GIS databases, and provide various dynamic analysis functions (Lin, 1993). Because GIS can provide many forest fire prevention and management functions, the Forestry Bureau has applied GIS in the following forest fire management situations:

(1) Forest fire danger rating warning system prior to the occurrence of fires. This system was completed in October 2002 and brought online for use. This system inputs the daily measured air temperature and fuel moisture (humidity) between 1,200 and 1,400 h into the database via the Internet. The server end estimates the values of forest fire danger-rating variables for each 1-km^2 grid. A logistical regression model is used to calculate the IP of every other grid and total the ratings after classification (Table 8.9). Finally, the forest fire danger-rating map is produced (Figure 8.9; Hsiao, 2003). The results are published on the website as a reference for on-site patrol personnel, who are instructed to pay attention when danger rating is high to prevent forest fires from occurring.

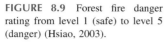

FIGURE 8.9 Forest fire danger rating from level 1 (safe) to level 5 (danger) (Hsiao, 2003).

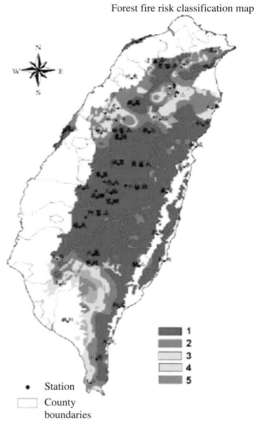

Forest fire risk classification map

8.1.6.1 Information Management of Forest Fire-Fighting Activities

To prevent forest fires from occurring, the Forestry Bureau established a forest fire danger rating warning system. Furthermore, the Forestry Bureau commissioned the Taiwan Forestry Research Institute in 2001 to introduce the forest fire, incident-command system (ICS) from the United States for the purpose of actively mobilizing personnel when forest fires occur, effectively managing fires, extinguishing fires in the shortest period possible, and reducing damages caused by forest fires. Personnel training during these few years, and operation during actual forest fires, has indeed improved in-efficiencies in past fire fighting work and has significantly shortened the amount of time in which forest fires can cause damage.

Forest fire ICS uses modernized geographical information technology, ICS concepts, and forest fire knowledge to build an information system that can be used as an auxiliary tool for extinguishing forest fires. ICS consists of a fire-reporting subsystem, a fire-status analysis subsystem, and a response-decision support subsystem (Ren, 2003).

The fire-reporting subsystem primarily reports real-time dynamic information of the fire site as the system's initial information, and uses GIS map calculation functions to rapidly determine the ignition point of the fire. This subsystem notifies the supervising unit, and provides a reference for fire fighter assembly points. The contents include the fire site and temporary command locations, on-site, fire-fighter deployment, the scope of the fire, and GIS-rendered graphical data.

The fire status analysis subsystem primarily compiles, analyses, and renders on-site information reported by fire fighters, forest fire weather data from mobile weather stations, and various digital maps related to fire-site information, for the purpose of understanding and grasping the fire-site situation. The fire-information subsystem functions include fire-site report records, depiction of the fire-site situation, maps of forest fire weather trends, and forest fire behavior predictions.

The response decision support subsystem is responsible for assigning fire fighting resources (request, reporting, deploying, and statistics), generating fire fighting meeting records, formulating fire fighting plans, and managing system files. This subsystem uses file management to store all reporting data and detailed, operating-process records into the system (request for resources, arrival of resources, deployment, and resource statistics). Fire fighting meeting records (fire fighting strategy and implementation) from the start of the forest fire (fire reporting, mobilization and assembling personnel, implementing fire fighting work, and current situation assessment) are made available for reading and output to provide commanders with real-time and accurate data as a reference for the chain of command.

8.1.6.2 Post-Forest-Fire Damage Assessment

Post-forest-fire processing primarily involves information collection, management, and analysis. The primary objectives are to rapidly restore vegetation and reduce secondary damage. In the past, post-forest-fire damage assessment of the burned (damaged) area by using manual labor was required to restore the forest as a reference. Individual trees in the burned area were surveyed to calculate the volume of wood and value as a basis for accountability and compensation. However, the steep mountain terrain of Taiwan often limits the transportation of personnel to fire sites for surveying; this results in inaccurate measurements. When forest fires extend over a wide area, the assessment of forest fire damage by this method confronts the limitations of manpower.

Due to advances in computer technology and the popularization of software and hardware equipment have enabled aerial and telemetric technology and GIS to be used to rapidly complete surveys of forest fire damage areas. When a forest fire has been extinguished, high-resolution images are recorded from aircraft. GIS coordinate recordings are used in coordination with other layer overlay references to render the forest fire damage area on the map (Figures 8.10 and 8.11). The damaged area is calculated, and forestry data (or natural forest type) for that area in the relational database is used to estimate the volume of forest damaged, and the damage value.

GIS provides a large amount of calculable data and technology, allowing post-forest-fire survey work to become faster and more convenient. This saves large amounts of manpower, as well as costs in conducting surveys, and provides diverse information as a reference for restoring forests.

8.1.7 Application of Satellite Telemetry, Multispectral Imaging to Forest Fire Severity Ratings

Lin et al. (2005) used ground survey data in coordination with SPOT satellite imaging to find the correlation between ground factors and NDVI, and developed an objective basis for rapidly assessing and defining the severity of forest fires using NDVI. Qiu et al. (2004) discovered that after removing interference sample points from the SR, NDVI, satellite constellation visualization, MSI, II, and normalized difference burn ratio (NDBR), all possessed the ability to determine the severity of fires. Of these, MSI, II, and NDBR (which possess midinfrared indicators and dual spectrums) exhibited superior determination functions.

FIGURE 8.10   Aerial photograph after a forest fire.

FIGURE 8.11   Aerial photograph after a forest fire crosslaid with other layers.

8.1.8 Research Content Regarding Taiwanese Forest Fires

Forest fires and climactic change research have become new foci in the field of forest fire research. Forest fires exert an important influence on global climate change and the environment.

8.1.9 Forest Fire Research Prospects

8.1.9.1 Application of 3S and Integration Technology

The development and direction of fire fighting work in various nations are moving toward 3S technology, which relies on digitizing forestry data, integrating forest fire warning, and monitoring system applications, and building automated, intelligent, and networked digital forest fire management systems. Forest fire research in Taiwan has also widely used 3S technology. In a study by Hsieh et al. (2011), detection technique changes were used to explore the use of high-resolution value aerial images on automated forest fire mapping. Liu (2004) integrated GIS and MODIS images for application to forest fire detection research. Chung and Chen (2002) integrated National Oceanic and Atmospheric Administration (NOAA) - Advanced Very High Resolution Radiometer (AVHRR) imaging with GIS to conduct forest fire monitoring research. These studies have all been centered around 3S technology for exploring forest fire warning and monitoring systems.

8.1.9.2 Forest Fire Prediction

Forest fires prediction combines weather factors, terrain, dryness of flammable items, types of flammable items, and ignition sources to analyze and predict the combustion risks of flammable items in the forest. Forest fire prediction has developed rapidly in various countries in the world since its inception in the 1920s. Taiwan's forestry department currently uses the study results of Hsiao (2003). Hsiao's study used a given day's highest temperature, temperature variation, accumulated period without rainfall, and drought index as weather factors to derive forest fire incident in a logistical regression model, and built a forest fire ignition probability model. Hsiao also considered space and time variations in weather factors, used GIS systems to conduct temperature and rainfall space-time estimates, and estimated forest fire hazard rating predictions for forests in Taiwan on a given day.

8.1.9.3 Forest Fire Management Policy and Promotion Research

Fire prevention and control must first consider human factors. Forest−protection policy is an essential part of social policy, and modern forestry is gradually shifting its focus to public fire-prevention education. Hong (2005) explored the relationship between forest fire prevention and fire-prevention promotion using the Dapu utility district in the Chiayi forest area, determined the optimal fire prevention promotional method, and established the optimal fire prevention promotion model. The research results indicated that the public perceived that rainfall and agricultural burning were the most influential factors that contributed to forest fires. People's perception of forest fire prevention and fighting importance were the highest in the arson penalties, updating fire fighting tools, burned forest, and reducing forest losses factors. Fire prevention promotion satisfaction levels were the highest in the television, poster, t-shirt, and fire-prevention drill factors. The surveyed people believed that the possession of fire-prevention knowledge by promotional personnel, the reporting of arsonists, and the cultivation of the love of forests were essential to promoting successful fire prevention. Legislation is one form of valuing forest fire prevention work. The use of policy and legislation to formulate suitable forest fire-prevention and fighting systems is a crucial topic. According to the Disaster Prevention and Protection Act, Article 3, Paragraph 6, the Council of Agriculture in the Executive Yuan is the central disaster prevention and response competent authority for forest fires, and the authority responsible for forests in Taiwan is the Forestry Bureau. Therefore, the prevention and fighting of forest fires are the primary functions of the Forestry Bureau. The complexity of forest fires also increases the difficulty of forest fire prevention. Thus, referencing the policies of advanced countries to search for an effective forest fire fighting strategy that is suitable to Taiwan to reduce losses is an urgent matter. Policy assessments assist the formulation, implementation, and feasibility of policies and programs, and allow resources to be saved to achieve maximum effectiveness. At the same time, the adoption of assessment rapidly allows expected goals to be achieved. Assessment results serve as a reference for revising programs and monitoring implementation personnel. Assessments can also significantly contribute toward international cooperation and implementing new coordinated policies.

8.1.9.4 Postfire Research

Forest fires consume forestry resources, and influence trees, vegetation, forest animals and plants, soil, and microbial growth. Forest fires also endanger the life and property of local residents. Fires also adversely influence the ecological system, and can damage ecological balance and cause the degradation of forest communities. Forest fires also cause air pollution. Assessments of fires must consider both economic loss and ecological effects. The assessment of forest fire damage and disaster levels must be objective. Forest fires in Taiwan included 406 fire alarms, 243 fires, 5,165 ha destroyed, costing an estimated NT$ 825,045,000. Because forest management requires long periods of time, the restoration of forests requires large capital and labor investments, as well as long waiting periods. Hsieh et al. (2003) used satellite images to assess forest fire losses. Hsieh primarily used satellite images in coordination with GIS to accurately calculate information related to the damaged area, the extent and scale of damages, and routes. Hsieh then considered historic forest fire cases, damage data, and causes of fires determined by previous researchers, natural factors (climate, humidity, and wind direction) to explore and assess the feasibility of building an estimation system.

8.1.9.5 Research Cooperation

For years, American and Canadian forest fire research personnel have implemented information sharing and academic exchange, and promoted two-way forest fire research development. At the same time, the United States and other countries such as Spain, Portugal, and Australia have engaged in long-term cooperation. This has facilitated an understanding of fire behavior, fire effects, vegetation dynamics, fire management strategy, the impact of socioeconomic factors on fire applications and fire fighting, and interactions between fires and global climate change. Possible future international cooperative studies include air quality management and understanding the regional and global impact of combustion products, assessing the influence of forest fire management strategy on the environment and benefits, large-scale monitoring and simulation of the influence of flammable vegetation cover management measures, flammable item mapping and monitoring methods and models for flammable item development and succession, and determining the mutual interaction between fires and other confounding factors.

8.1.9.6 Aerial Forest Preservation

At the same time as forest fire prevention plans are implemented, improving fire fighting ability has become a crucial topic for forest fire prevention organizations in various countries. The 1940s were the first stage of aerial forest preservation; fire fighting in the 1940s primarily involved using airplanes to patrol, observe, and report fires. After the 1950s, countries with large forest areas generally developed aerial forest preservation, and began using airplanes to directly extinguish fires. The Taiwanese government first used the air fire brigade reserve from the National Fire Agency to directly fight fires in the 2002 Lishan forest fire. On March 10, 2004, the National Airborne Service Corps reserve of the Ministry of the Interior (hereafter referred to as the airborne corps) was formed, which integrated the Aerial Police Brigade of the National Police Agency, the air fire brigade from the National Fire Agency, the General Aviation Squadron of the Civil Aeronautical Administration (Ministry of Transportation and Communications), and the official aerial fleet of the Coast Guard Administration's aerial reconnaissance corps. The airborne corps is responsible for coordinating service scheduling and providing support during land and sea disasters, rescue, medical aid, observation and patrol, and transportation tasks. The airborne corps cooperates closely with forestry units and fire fighting units in implementing various kinds of disaster relief training. They have managed multiple forest fires in recent years, and have used helicopters to directly suppress forest fires with satisfactory results (Chen, 2009). Current aerial forest preservation research primarily involves satellite forest fire monitoring by using plane-mounted infrared detection of ignition sources. Plane-mounted infrared can be used for conducting planned continuous tracking and monitoring of forest fire extensions. This enables various types of real-time dynamic information in major forest fire sites to be accurately grasped.

8.2 CONCLUSION

Climatic changes in recent years have caused serious forest fires to occur in many countries. Worldwide, this poses a significant challenge to forest fire prevention, and reform must be undertaken. Because forest fires do not occur frequently in Taiwan, the technical research and development behind forest fire prevention in Taiwan is lacking compared with the intermediate and long-term development observed in other countries. However, forest fires are extremely crucial in protecting forests, and preventive measures are one of the best solutions for managing forest fires. This highlights the importance of forest fire danger rating warning systems. However, forest fires in Taiwan are mostly regional forest fires, and human factors are the primary cause of fires. If patrols can be improved for community forestry and forest fire prevention education can be promoted, the fire occurrence rate can be reduced.

When forest fires occur in Taiwan, crisis management must include a comprehensive fire fighting plan. Postfire damage assessments and forest restoration all require the following forest fire-related research: forest fire detection technology, forest fire fighting strategy in complex terrain and weather conditions, relevant fuel moisture content forecasting and monitoring technology, and fire behavior models.

REFERENCES

Abhineet, J., Shirish, A.R., Singh, R.K., Das, K.K., Roy, P.S., 1996. Forest fire risk modelling using remote sensing and geographic information system. Current Science 70 (10), 928–933.

Agee, J.K., Pickford, S.G., 1985. Vegetation and fuel mapping of North Cascades National Park. Final Report. College of Forest Resources. University of Washington, Seattle, Wash.

Anderson, H.E., 1982. Aids to determining fuel models for estimating fire behavior. Intermountain Forest and Range Experiment Station. USDA Forest Service, Ogden, UT. General Technical Report INT-122, April. 28 p.

Cherng-Ming Wang. 2002. A study on the prevention of Forest fires and the Establishment of Firebreaks. The case of Forest conservation and management administration. Theses and dissertations of National Taipei University of Technology, pp 102.

Ching-Sheng Chang. 2004. A study on Characteristics of fire-caused factors and strategies of fire-fighting mechanism—a case study of historical data 1989–2003 in Chiayi Forest district. Theses and dissertations of National Taiwan University, pp 88.

Chi-Wen Hsiao. 2003. The Study on Estimating Temporal-spatial Distribution of Forest Fire Danger Rating in Taiwan. Theses and dissertations of National Taiwan University, pp 83.

Chiou, Chyi-Rong, Tseng, Jen-Chien, Huang, Wen-Dar, Yang, Chi-Ming, 2004. Grey Relational Analysis of the Effect of Climate Factors on the Satellite Remote Sensing Brightness Index (BRI) of Guineagrass in Mt. Dadu Area. Crop. Environment & Bioinformatics 1, 207–214.

Chen, Chorng-Shyan, 2009. Helicopter Utilization on Forest-fire Relief. J. Taiwan Disastr Prevent. Soc. 1 (1), 40–44.

Chung, Yuh- Lurng, Chen, Cheng-Hua, 2002. Study on integrate NOAA-AVHRR image and geographic information system to monitor Forest fire. J. Photogramm. Remote Sensing 7 (2), 87–96.

Chung, Y.L., Chen, C.T., His, C.N., Liu, S.M., 2006. Establish on MODIS Image into Drought Monitoring Model in Taiwan. J. Photogramm. Remote Sensing 11 (1), 39–58.

Chuvieco, E., Congalton, R., 1989. Application of Remote Sensing and Geographic Information System to Forest Fire Hazard Mapping. Remote Sensing of Environment 29, 147–159.

Chuvieco, E., Salas, J., 1996. Mapping the spatial distribution of forest fire danger using GIS. International Journal Geographical Information Systems 10 (3), 333–345.

Council of agriculture executive yuan forestry bureau, 2000. Forestry statistics in taiwan. Council of Agriculture Executive Yuan forestry Bureau Press, Taipei Taiwan.

Fuller, M., 1991. Forest Fires: an introduction to wildland fire behavior, management, firefighting and prevention. John Wiley and Sons Inc., New York.

Hsieh, Li-Cheng, Luan, Jar-Miin, Yen, Tian-Ming, 2004. Study of the Application and Loss Area Evaluation from Natural Disaster Using Satellite Image-Example on Forest Fire. Ministry of Science and Technology RGB. Ministry of Science and Technology RGB, PG9302–0611.

Hsieh, Y.T., Chung, Y.L., Liao, C.S., Yui, Y.G., Teng, K.C., Wu, S.T., 2011. Application On High Resolution Digital Aerial Images for Automatic Forest Fires Mapping by Change Detection Techniques. J. Photogramm. Remote Sensing 16 (1), 11−22.

Hsien-Ren Chang. 1999. Practical Problems of Forest Fire Prevention and Suppression—a Case Study of Taiwan Forestry Bureau. Theses and dissertations of National Chung Hsing University, pp 68.

Hwang, C.Y., Lin, C.C., 2005. Analysis of Forest Fires in Taiwan National Forests. Quarterly Journal of Chinese Forestry 38 (4), 449−464.

Jiun-hau Ren. 2003. The Study of the Establishment of Incident Information System for Forest Fire. Theses and dissertations of National Taiwan University, pp 129.

Lin, Chau-Chin, 1992. The Development of Forest Fire Database for Taiwan. Quarterly Jounal of Chinese Forestry 25 (3), 63−72.

Lin, Chau-Chin, 1993. Forest fires and firebreaks of Ta-Chia-His national forest. Taiwan J. Forest Sci. 8 (2), 159−167.

Lin, Chau-Chin, 1994. Study on Rating Fire Risk Zone of National Forests of Taiwan. Bull. Taiwan For. Res. Inst New series 9 (1), 61−72.

Lin, Chau-Chin, 1995. Study on the predicting system of forest fire danger rating in Taiwan. Taiwan J. Forest Sci. 10 (3), 325−330.

Lin, Chau-Chin, 2001. Forest Fire Management Strategy in Taiwan from the United States Forest Fire Season at millennium. Taiwan Forestry Journal 27 (1), 13−17.

Liu, 2004 Chen-Jung Liu. 2004. Investigation of primary vegetation on coastal Windbreak Burned Site in Taichung Harbor. Theses and Dissertations of National Chung Hsing University, pp 73.

Lin, Chau-Chin, Chiou, Chyi-Rong, Chou, Chiao-Ying, 2005. Identifying and evaluating fire severity: a case study of the Wulin fire. Taiwan J. Forest Sci. 20 (3), 203−213.

Richard, K.M., van Lear, D.H., 1998. Hurricane−fire interactions in coastal forests of the south: a review and hypothesis. Forest Ecol. Manag. 103 (2−3), 265−276.

Shih-Ming Liu. 2004. Study on integrating GIS techniques and MODIS image into Forest fire detection. Theses and Dissertations of National Pingtung University of Science and Technology, pp 92.

Shu Hsia Hong. 2005. The Study of Apply Forestry Extension on Forest Fire Prevention. Theses and dissertations of National Chi Yi University, pp 99.

Vakalis, D., Sarimveis, H., Kiranoudis, C., Alexandridis, A., Bafas, G., 2004. A GIS based operational system for wildland fire crisis management I. Mathematical modelling and simulation 28 (4), 389−410.

Yang, M.H., Gao, Y.Y., E, M.S., 1992. Development and Application of the Potential Compartment Forest Fire Danger Grade Map. Scientia Silvae Sinicae 28 (6), 532−537.

Yen, Tian-Ming, Wu, Jing-Yang, 2004. Influencing factors of Forest fires in Nantou Forest district. Quart. J. Forest Res. 26 (1), 47−60.

Chapter 9

Wildfires in India: Tools and Hazards

Joachim Schmerbeck

TERI University Department of Natural Resources, Vasant Kunj, New Delhi, India

Daniel Kraus

European Forest Institute (EFI), EFICENT Regional Office, Freiburg, Germany

ABSTRACT

Fires, set by humans, have both positive and negative outcomes. Throughout India, this practice has been followed for thousands of years, thereby shaping the Indian landscape over the years. This chapter describes the many reasons, as far as they are now known, why people in India set fire to the landscape, and to what extent these fires are hazardous. Review shows that large destructive fires are infrequent in India, most being low-intensity fires that burn only the ground vegetation in their respective ecosystems. Over the years, fires have had a strong impact on ecosystems and their functioning. A study of the different reasons why people burn forests reveals one and perhaps the most important reason is that fire is essential for people's livelihood. This chapter also highlights that the way in which fire is used and its role in managing segments of the landscape are not in accordance with the current forest fire policy. We suggest a more flexible approach that allows for the inclusion of fire as a tool in local fire management systems.

9.1 INTRODUCTION

India's landscape as it appears today is to a large degree a product of extensive fires. Reports of fires in India first appear in historical records, but fires have been applied to landscapes long before that (Pyne, 1994). Fires help determine forest structure and composition, alter soil properties (Shakesby, 2011), and contribute to air pollution (Singh and Panigrahy, 2011). However, our understanding of Indian wildfires is superficial. We know that the majority of fires are caused by humans (Bahuguna and Upadhyay, 2002; Pyne, 1994) and that the motivation for setting them varies across the country (see below). However, the proportional breakdown of the different causes of wildfires in India is unknown due to lack of empirical assessments. Nevertheless, it is emerging that these fires

Wildfire Hazards, Risks, and Disasters. http://dx.doi.org/10.1016/B978-0-12-410434-1.00009-9
167

are important for local livelihoods as they make available many important domestic and commercial goods Schmerbeck et al. (in press). Lack of empirical data leaves gaps in the understanding of issues such as how many of the 275 million people (estimated by the World Bank (2006)) rely on the burning of India's forests, how much area needs to be burned to satisfy the requirements of forest dwellers, and how much more area gets burned beyond their needs.

In India, two fire management practices are followed: First, the British Raj implemented the official regime constituting the "ban and punishment policy," and the second involves a need-based application of fire by several forest-dwelling groups. This contradictory policy has been a source of conflict since its inception in the British Raj leading to resistance of this fire-suppression ordinance in many parts of India, christening it "oppression of the people" (Pyne, 1994). Over millennia, forest dwellers have depended on landscapes using fire as a tool to gain benefits from cultivated and uncultivated land. Burning modifies vegetation for a short term, for instance, grasses suitable for livestock grazing resprout after fire. A change of species composition from fire tolerant to fire intolerant is prevented by the regular application of fire. But regular burning also precludes the services that accrue from advanced succes-sional stages of vegetation.

This chapter provides an overview of fires in India: those that are used as a tool and those that are hazardous. We begin with a section on the history of fire and what is presently known about the fire regime and continue with a general outline of the ecological impacts wildfires have in India. This is followed by a section on the utilization of fire as a tool in India's forests and on the hazardous effects of fires in India. To conclude, we provide an outlook on the research needs and management implications that we see in this context.

9.2 FIRE HISTORY AND REGIMES IN INDIA

For millions of years, lightning fires have driven the evolution of fire-adapted traits in plants (Keeley et al., 2011). Today, large areas occupied by flammable biomes in the tropics and subtropics are mostly attributed to anthropogenic burning. Yet the occurrence of natural fires and flammable ecosystems predate anthropogenic burning by millions of years. Recent evidence shows that the expansion of C_4 grasslands during the late Miocene was driven by natural fires rather than by decreasing atmospheric CO_2 (Keeley and Rundel, 2005). We know from palynofossil records that fire was present in south Asian forests and grasslands for at least the last 20 million years (Mathur, 1984). This is also reflected in the Pleistocene faunal records from India indicating that envi-ronmental conditions favoring fire as an important factor to shape habitats existed then (Badgley, 1984).

There is no way to prove for how long India's forests have been exposed to anthropogenic fires. Gadgil and Meher-Homji (1985) (as cited in Saha (2002), p. 1) think that humans have been burning forests in India for 50,000 years, while Goldammer (1993) reports that the dry deciduous forests of Asia have been affected by human activities, including human use of fire for about 12,000 years. Misra (1983) pointed out that India's tropical, subhumid, and dry deciduous forests, which once covered vast areas of the country, have been almost entirely replaced by savannas due to human-influenced fire regimes. Thus, the anthropogenic modification of these ecosystems has been ongoing for a long time (Bor, 1938; Karanth et al., 2006; Noble, 1967; Thomas and Palmer, 2007).

Whenever people moved, they took fire with them. After the Aryan invasion of India, around 2000 BCE, pressure on the forest slowly began to increase. This increase according to Stebbing (1922) was the result of constant invasions by peoples from central Asia who brought their herds of cattle with them. To feed the herds, they slashed and burned forests to create grazing grounds. This land conversion process peaked under the reign of the Muslims around AD 800. The original land owners were driven into the remaining forests (Stebbing, 1922) where they practiced shifting cultivation and used the forest as a source of raw materials.

It is believed that the contemporary crisis in forest and fire management is profoundly embedded in historical processes through which state forestry institutions have evolved (Chakraborty, 1994; Poffenberger and McGean, 1996). Regular forest management started in 1864 with the appointment of Dietrich Brandis, a botanist from Germany, as the first Inspector General. Following the passing of the first Forest Act in 1862, Brandis' main objectives were to demarcate and limit entry to forests reserved for the government, to protect these forests from fire, illegal felling and grazing, and to organize the forest administration (Hesmer, 1986; Mammen, 1964).

The dynamics of fire regimes in India today are largely a consequence of imperial expansion and the perception of fire under the British Raj (Pyne, 1995; Rakyutitham, 2000). Rakyutitham (2000) maintains that conventional views of environmental problems are influenced and defined by powerful groups within a society, and such views usually become unquestioned "knowledge." This was certainly the case during the time of the British Raj, which introduced European fire practices as an inextricable part of colonial enterprise. Before the arrival of the British, the practice of land burning was deeply embedded in India. Both Indian culture and society depended on the benefits of burning (Pyne, 1995). The British foresters in India saw fire as a destructive agent that reduced state revenue from forests and destroyed property and therefore had to be suppressed and excluded (Pyne, 1997). For colonial authorities, fire was not only an element that was difficult to control but also one that had to be controlled in a "civilized" society. Consequently, the Raj banned agricultural practices involving fire (Pyne, 1995).

Local people opposed to the commercially oriented colonial policy continued to practice the annual burning of forest as a form of protest and as an attempt to restore customary practices to their traditional lands. The role of the community versus that of the state in managing forest resources has been debated for as long as local people and the state have been depending on forests (Poffenberger and McGean, 1996; Springate-Baginski and Blaikie, 2007).

As the forest area declined, British foresters began to realize at the start of the twentieth century that too much fire protection had changed the forest structure and composition as fire had been maintaining it. They recognized that "fire had not been random and ravenous… but was applied to particular sites at particular seasons for particular purposes and for particular people" (Pyne, 1994, p. 13). In 1914, fire was formally reintroduced as a forest management practice in India, and British foresters tried to reintroduce fire into the landscape. They launched an elaborate fire management program including fire lines, block lines and guidelines, and early clearing and burning. However, historical fire regimes could not be restored.

9.3 FIRE AND ECOLOGY

In large parts of India, the landscapes consisting of tropical and subtropical fire-prone savannas and woodlands have been shaped by a long history of human-influenced fire regimes. In the seasonal tropics, unlike the humid or moist tropics, fire is an ecologically important disturbance factor. The dry deciduous forests and savanna grasslands of the Indian subcontinent have coevolved with fire and many of the plants have adapted to fire. Although fire impacts are widely recognized as important ecological factors in some of India's ecosystems, many ecologists still fail to incorporate the role of fire into their thinking about natural ecosystems (Keeley and Bond, 1999, 2001; Saha and Howe, 2001). Fire has a strong ecological impact on organisms directly (Bond and van Wilgen, 1996) and indirectly, for example, by changing soil properties (Mataix-Solera et al., 2011; Shakesby, 2011; Shakesby et al., 2007) and hydrological cycles. A sound understanding of the fire ecology in south Asian ecosystems is a prerequisite to the identification of social and ecological constraints and benefits of fire. So far, there are no studies available that allow us to reconstruct how ecosystems in India would have looked without human-caused fires burning over wide areas over a long time. We can only guess the structure and functioning of ecosystems under a scenario without fires.

Fire-dependent ecosystems are those where fire is an essential process in maintaining species composition, biodiversity, and structure. They are often referred to as fire-adapted or fire-maintained ecosystems. Frequent fires maintain fire-dependent vegetation formations by selecting for fire-tolerant species (Furley et al., 2008) such as those that are able to coppice (Bond and van Wilgen, 1996), mainly from meristems below the soil surface (Saha and Howe, 2003). These formations are characterized by a wide grass/tree ratio and an abundance of C_4 grasses (Ratnam et al., 2011; Veldman et al., 2012) and other light-demanding plant species. A change in the fire regime, for example, due to fire suppression, "releases ecosystems" from their "arrested states" (Favier et al., 2004; Hopkins, 1992) allowing for an increase in the proportion of woody species. Under tree canopies, C_4 grasses get shaded out and are replaced by C_3 grasses, and eventually, shade-tolerant tree species establish (Ratnam et al., 2011) if their propagules reach the site. In the absence of fire, and where soil properties and precipitation allow, succession may convert savanna and open forests into closed forests.

Examples of fire-dependent ecosystems can be found throughout south Asia. It is apparent that most of the open forests, savannas, and grasslands found in India today have developed as a consequence of human-influenced fire regimes. These new regimes have replaced the different forms of closed forest formations that once existed (Blasco, 1983; Cole, 1986; Goldammer, 1993; Misra, 1983). Today, <3 percent of India's land area is occupied by forests with >70 percent crown cover (Forest Survey of India (FSI), 2011) and half of all these forests are considered to be fire-prone (FSI, 1995). The fire susceptibility varies greatly between the different states (33 percent in West Bengal to 93 percent in Arunachal Pradesh) (Government of India, 1999). However, studies covering more than one year on dry forest ecosystems in India suggest an increase in tree diversity over time (Puyravaud et al., 1995; Saha, 2003; Saha and Howe, 2003).

The fire-influenced dynamics described above, between open grass-dominated systems and closed tree-canopy systems, have altered the abundance of large herbivores and predators dependent on them (Landsberg and Lehmkuhl, 1995). As long as grass-dominated ecosystems are maintained by fire, herbivores will have abundant forage (Johnsingh, 1986; Main and Richardson, 2002; Tomor and Owen-Smith, 2002). After a fire, the number of herbivores increases due to the higher nutrient levels available in the herb layer (Carlson et al., 1993; Moe and Wegge, 1997). However, in India, fire is employed to conserve the habitats of large herbivores, for example, the one-horned rhinoceros (*Rhinoceros unicornis*) and predators like tigers (*Panthera tigris tigris*) (Landsberg and Lehmkuhl, 1995). These fires are even set in areas where fires would be very unlikely without human ignition, for example, the Manas National Park (Takahata et al., 2010).

9.4 FIRE AS TOOL

The answer to the question "Why do people set fires to the land?", is more or less known in India. However, detailed studies show that unknown motivations to set fire exist within different cultural and ecological contexts, giving a fair idea as to what motivates people to set fires, but we have only a vague idea as to how many fires there are on a national or local scale. Regarding the motivations to apply fire to a particular landscape, Pyne (1994) refers to observations made by a forest officer in the Ghumsur Forest in Orissa at the beginning of the twentieth century who held bamboo cutters and forest license holders responsible for many of the fires. Additionally, fires helped to smoke bees out of trees for their honey, burn under Mango and Mohwa trees (also Mahwa or Mahua (*Madhuca longifolia*)) to clear the forest floor (to facilitate the finding of fallen fruits and flowers), roast tree seeds, burn the undergrowth around villagers (to increase visibility and reduce cover for tigers and panthers), clear land, produce better quality pastures, and to drive game out of the forest for hunting. A more quantitative observation was made at around the same time by the forest administration of the Madras presidency for 1922−1923 (Forest Department Madras Presidency, 1923), which stated that of almost a third of all forest fires started outside of the Reserved Forest, one-third could not be explained. Almost half of the remaining third were caused by hunters using fire to drive game and by graziers burning pastures to produce new grass shoots. Uncertainties that exist in the historical forest records mean that a large fraction of fires cannot be allocated to a cause. A further problem with the records is that they fail to provide reliable data about the total number of fires. Information regarding fire occurrence is obtained from Moderate Resolution Imaging Spectroradiometer (MODIS) data, but it can be expected to miss a great number of fires (Hawbaker et al., 2008).

However, from what is known today, the reasons why fires are lit in India can be allocated to six broad categories: (1) cultural, (2) utilization of nonwood forest products (NWFPs), (3) collection of wood products; (4) accessibility and safety, (5) pasture (land clearing) and hunting (game drives), and (6) agriculture.

9.4.1 Culture

Local people from Rajasthan's Satukonda Wildlife Sanctuary and Kumbalgarth National Park state that fires are lit annually to worship their Gods, with fires forming a golden necklace to worship the deity. Such observations made by the authors in the field have not been documented in detail for India until now. However, Krishna and Reddy (2012) state that "ethnic belief of tribes to worship the God" is the main source of wildfires in Rajasthan.

The use of fire is something that is inherited in India, and it has been so since the beginning of Indian culture (Pyne, 1994). Among some rural people there is also the belief that fire will initiate rain. This belief has been reported from Mexico by Goldammer (1978), and when local people in south India were asked for the reasons behind the annual fires in a dry forest (Schmerbeck, 2003), they gave the same answer. Roveta (2008) studied the use of fire among members of the Soligar tribe of the Biligiriranga Swamy Temple Wildlife Sanctuary, and learned that the burning of the forest floor for ceremonies and recreational purposes accounted for 11 percent of the importance fire had among other purposes.

However, considering the broader picture, it is felt that a much stronger motivation for setting fire to a particular part of the landscape exists as the potential for tangible gain.

9.4.2 Nonwood Forest Products

Since the hunter-gatherer times, NWFPs or forest products not obtained by the harvest of trees, have been important to humans. In this chapter, the term NWFP will be used according to Wickens (1994), who describes it as "all the biological material (other than wood products as defined above) that can be utilized within the household, be marketed or have social, cultural or religious significance" (p. 56). However, this definition does not include forage.

In India, for people who depend on forests, NWFP often plays a key role in their livelihoods (Hunter, 1981; Yadav and Dugaya, 2013). The provision of many of these products is strongly linked with fire because the products are obtained from forest ecosystems in which fire plays an important role. Case in point is the Tendu tree (*Diospyros melanoxylon*), a deciduous tree occurring in dry forest of the Indian Deccan, the leaves of which (Tendu Patta) are used for the production of small cigarettes (*bidis*) in India. This is the most well-known example of a commercial NWFP requiring fire for its production. The trees used for the Tendu Patta production are shrubby, and this appearance is a result of regular burning, which keeps the tree from growing to its full size. Fire also triggers the flushing of new leaves that are collected while they are still in an early stage of development. The *bidi* business is large, but just how large depends on the source of the figures. According to Lal (2012), 7.5 million people are employed part time to pluck tendu leaves, and 4.4 million women and children are engaged in manufacturing these cigarettes. In 2002–2003, the World Bank (2006) estimated that 360,000 ton of tendu leaves were produced, which is around 100,000–200,000 ton less than that in the previous two decades (1980–2000). But these figures are uncertain because it is difficult to obtain data for tendu and *bidi* production (Lal, 2012). Nevertheless, the tendu and *bidi* business give an indication of the extent to which products tied to fire cause an impact on the livelihood of rural people. There are several such examples to understand this connection between fires, products, and livelihoods; however, they are not all as commercially important as tendu leaves are. Table 9.1 provides a list of known NWFPs to date in India that are produced, utilized, or benefit from the use of fire. The fact that there are only a limited number of studies that have been done on this topic makes it likely that there will be more NWFPs added to the list in the future.

The importance of NWFP for the livelihoods of forest dwellers varies from location to location. Schmerbeck et al. (in press) found that there was a strong variation in the importance of forest products between villages in one area of Andhra Pradesh, and that the NWFPs that were made available by regular forest fires served mainly domestic purposes.

TABLE 9.1 The Role of Fire in Producing and Utilizing Nonwood Forest Products (NWFPs) from Indian Forests

NWFP	Species	The Role of Fire	References
Tendu leaves	*Diospyros melanoxylon*	Maintaining shape of plants and enhancing leaf flush	Hunter (1981), Saigal (1990), Goldammer (1993), Yadav and Dugaya (2013)
Honey	Miscellaneous	Induction of flowering and the production of smoke to drive away bees	Schmerbeck et al. (in press), Schmerbeck (2003)
Roofing material	Mainly C$_4$ grasses (e.g., *Themeda cymbaria, Cymbopogon* spec.)	Regular fires maintain abundance of required grasses	Roveta (2008)
Amla fruits	*Phyllanthus emblica*	Control of hemiparasites and enhancing production	Kohli (2010), Ganesan and Setty (2004), Rist et al. (2010)
Mahua seeds and flowers	*Madhuca longifolia* (syn. *Madhuca indica*)	Fire is lit prior to flowering/fruit ripening. The cleared and blackened ground improves access and the visibility of fruit	Saigal (1990), Pyne (1994), Nanda and Sutar (2003)
Tree seeds	*Terminalia chebula Buchanania lanzan*	Enhancing production	Kohli (2010)
Brooms	Several grass species (e.g., *Aristida setacea*)	Regular fires maintain abundance of required grasses	Kohli (2010), Roveta (2008)
Eatable tubers and roots	Miscellaneous	Fire improves growing conditions for the species and makes the plants accessible	Kohli (2010), Roveta (2008)
Fertilizer	All	Postfire rains transport ash to agricultural fields	Pyne (1994), Roveta (2008)
Leaf cup	Unidentified	Fire triggers the flush of leaves	Kohli (2010)

9.4.3 Wood Utilization

The role of fire in timber production has been reported from colonial times in India. Pyne (1994) provides a nice overview of the position held by the forest administration regarding the role of fire in forest management. The British administration applied the European forest protection policy to the Indian forest with the goal of protecting several timber species (mainly Sal, Teak, and Chir Pine stands) as well as commercial bamboo from fire. This policy thoroughly changed the ecology of the ecosystems involved and made the regeneration of the targeted species uncertain. Even though the debate regarding the role of fire never led to a clear consensus, it became evident that fire was required for the continuity of many species on a given site. However, today timber production is not a priority in India so the use of fire plays a minor role in the debate about forest fires.

Much more important and under estimated by scientists, ecologists, and land managers in India is the role of fire in producing fuel wood. This connection has been observed in a few local studies (Schmerbeck et al. (in press); Roveta, 2008; Schmerbeck, 2003) and has been referred to in some studies outside India (e.g., Hough, 1993 (Benin); Vayda, 1996 (Indonesia); Laris, 2002 (Malawi)). Light-to-medium-intensity fires fully or partially kill the above-ground parts of woody plants. Because the majority of these fires are of a low intensity, the wood is not fully burned and leaves behind dry wood that can be collected for fuel wood. Fire also clears the ground and makes the area more accessible for fuel wood collection. Such practices involving fire can be improved through regulation by the forest administration, which manages the collection of dead wood (Schmerbeck, 2003). Also observed by Schmerbeck (2003) was the collection of bark from *Euphorbia antiquorum*, a tree-like xerophytic plant, after it had been killed by fire. The bark was used to produce charcoal locally. The charcoal was sold in a nearby town to goldsmiths who valued this fuel for its steady flame.

9.4.4 Accessibility and Safety

Pyne (1994) again provides us with an eye-witness account of fire used to make an area safe from dangerous animals. He describes a practice used by tribes during the British Raj: "A veteran Conservator of Forests, G.F. Pearson, noted that even the Ghonds, a long-enduring tribe of Indian central forests, 'never go into the jungle now, where tigers are supposed to live, without setting it on fire before them, so as to see their way'" (pp. 3, 4). However, fire does not only provide safety from large predators but also safety from poisonous vertebrates and invertebrates (e.g., snakes, scorpions, and leaches) as well as insects like ticks (Roveta, 2008).

Although it cannot be empirically proved, it seems likely that keeping the landscape, especially the forested parts, easily accessible is more important to the people than making an area safe; nonetheless, forest clearance achieves both objectives. Roveta (2008) quotes a respondent to a survey who stated "I was walking in the forest with my father—at that time I was very young—and the path was closed by the bushes, therefore my father decided to set fire in

FIGURE 9.1 A degraded dry forest in Tamil Nadu, South India: wildfire provides accessibility by reducing thorny woody species and grasses. *Photograph courtesy: Joachim Schmerbeck.*

order to clean the way" (p. 37). It is just such human behavior (making a decision to burn) that caused the growth of such pioneering, and thus the conditions for the growth of woody thorny bushes and tall grasses in the first place. Both these species make it difficult to move through the forest, while after a fire, it becomes very easy (Figure 9.1).

Access improvements also allow herders to more easily move their livestock from one place to the other (Roveta, 2008). The close connection that exists between accessibility and use of forest products in the context of wildfires is illustrated by Kohli (2010). She investigated the utilization of tangible ecosystem services that depend in one way or the other on the presence of wildfire. She surveyed 557 households in 14 villages adjacent to a dry forest in Andhra Pradesh. Respondents were asked to rank the importance of ecosystem services associated with wildfire. Highly ranked was the accessibility to the forest due to burning, while at the same time there were some household respondents who did not even mention accessibility.

9.4.5 Pasture and Hunting

The practice of setting fire to vegetation to improve forage is a worldwide practice. The principle is straightforward: forage species, mainly grasses, are set to fire, but the grasses are not entirely consumed. Fire removes the parts of the grass that are above the soil surface, and often nothing but the bare soil remains. The ash fertilizes the soil leading to a fresh flush of grasses (Laris, 2002). These fresh grasses, due to their higher nitrogen (Lü et al., 2012) and crude protein content (Mbatha and Ward, 2010), are a better quality fodder than are grasses that sprout without being burned. Based on the frequency with which fires are mentioned in the literature in connection with forage improvement (Hough, 1993; Kepe and Scoones, 1999; Laris, 2002; Mistry et al., 2005; Shaffer, 2010; Vayda, 1996), this use of fire is likely predominant worldwide. In India, fires set to enhance the availability of fodder for domestic animals are the most common cause of wildfires (Brandis, 1897/ 1994; Goldammer, 1993; Government of India, 1999; Kohli, 2010; Sinha and Brault, 2005; Schmerbeck and Seeland, 2007; Roveta, 2008). However, none of these studies looked at the amount of area burned compared to the actual requirements for forage ground or fodder.

The effect that fire has on the quality of resprouting grasses also attracts wild herbivores. Hunters use fire in this way to attract prey. Further, poachers have been known to use this same tactic, although much less land gets burned in this manner. The early inhabitants of the country also used fire for hunting (Goldammer, 1993; Government of India, 1999). Fire is used today to provide habitat for herbivore species targeted in biodiversity conservation approaches (Takahata et al., 2010). It has been reported that fire is used by Australian Aborigines (Gott, 2005) to drive prey out of thick vegetation into the open where it can be killed. Although, there are no recent reports from India about the use of such sophisticated practices, it is likely that they did occur in India as historic records allocate a significant proportion of the fires to this reason (Forest Department Madras Presidency, 1923).

9.4.6 Agriculture

A look at a fire distribution map of India (Vadrevu et al., 2013) reveals two main wildfire hotspots: the north western plains and the forests of the northeast. The cause of the wildfires in the northwest has been linked to an agricultural practice: that of setting fire to postharvest fields (Kharol et al., 2012). In the northeast, wildfires are mainly caused by slash-and-burn agricultural practices in forests (Vadrevu et al., 2013). The burning of postharvest fields and slash-and-burn agriculture are also practiced elsewhere in India, but nowhere are these as frequent as in these two regions.

Burning in the agricultural sector has often been reported as a cause of forest fires. In fact, a large proportion of the total wildfires in India are caused by farmers. This happens when postharvest fires escape and end up burning forests (Forest Department Madras Presidency, 1923; Kurian and Singh, 1996; Roveta, 2008) and also when clearance fires, used for clearing land for agriculture, escape into forests (Bahuguna and Upadhyay, 2002; Government of India, 1999). When fire is applied to agricultural fields, it is considered a tool but when it gets out of control (escapes) and starts a forest fire, it is an accident. The figures available for such accidents range from 40 percent (Vadrevu et al., 2008) to 69 percent (Singh and Panigrahy, 2011) of the total fires in India annually. Fires have a strong impact on air quality

(Kharol et al., 2012; Singh and Panigrahy, 2011) and may influence agricultural soils (Mataix-Solera et al., 2011) and water regimes.

In the northeast of India, fires are mainly caused by shifting cultivation carried out in forests. In India, the practice is known as "Jhum," (Semwal et al., 2003). Kingwell-Banham and Fuller (2012) reviewed the practice of Jhum in India and Sri Lanka and identified five ethnic groups, based on language families that were practicing Jhum. Most of the groups from India were in the northeast, the northeast peninsula, and in the southern part of the Western Ghats. According to these authors, the groups practicing Jhum are a mix of three different strategic groups. Only one of these groups consists solely of shifting cultivators, while the other two groups also consist of hunter-gathers and settled agriculturalists. However, it is in the north east region of India where Jhum is clearly most intensively practiced judging by the number of the fires (Vadrevu et al., 2013) and the number of groups practicing it (Kingwell-Banham and Fuller, 2012). However, the number of people demanding land for shifting cultivation in the northeast is increasing. The increasing number of people and their land-use practices result in some serious consequences such as a loss of soil fertility, decreasing forest cover, a reduction in market crop productivity (due to shorter agricultural cycles), reduced system stability and resilience, loss of biodiversity, and large-scale desertification (Ramakrishnan, 2007).

9.5 FIRE AS A HAZARD

A common definition of natural hazards is the threat of a naturally occurring event with a negative effect on people or the environment (Wisner et al., 2004). During the annual long and intense dry seasons in the region, fires are a regular phenomenon, and many of them have the potential to cause major damage. Most fires in India do not directly threaten lives and property, but they may have a strong impact on human health by way of the haze and smoke they produce. However, some fires do directly affect lives and property even though they are rare. By far, the most serious negative effect of such hazardous fires in India includes the loss of livelihood for tribal and rural poor people. Additionally, they result in a loss of timber, NTFPs, fuel wood, fodder, and soil fertility. They also degrade catchment areas when a certain threshold for fire occurrence and frequency is reached. Further, they damage cultural heritage sites and the land-use systems that provide the basis for the livelihood of forest-dependent people living in rural India (Poffenberger and McGean, 1996). Hazardous fires that burn sensitive mountain ecosystems can be the indirect cause of landslides, mudslides, erosion, increased water run-off, flash floods, and soil depletion. Although official statistics and information on fire losses are rather weak, the costs to regenerate forests damaged by forest fires are estimated to be Rs 4,400 million (US$96 million) annually (Bahuguna and Upadhyay, 2002).

Regular fire use for land-use activities increases the probability of uncontrolled fires. Some of the affected ecosystems are extremely sensitive to fire, but without subsequent ignitions, these ecosystems can recover (Cochrane, 2002; Myers, 2006). An overall trend of increased ignitions and the consequent dramatic changes in vegetation structure and fuel characteristics can be observed. This is especially the case where fire creates a positive feedback loop that leads to increasing flammability and drier conditions even in nonfire-prone vegetation (Cochrane, 2001; Goldammer and Mueller-Dombois, 1990). The excessive use of fire in association with an increase in population and land-use changes has led to a shift in vegetation types toward more pyrophytic (tolerate fire) plants as described above. The impact of these changes is strongly dependent on the social and ecological effects of contemporary fires. Despite the existence of traditional fire management systems such as Jhum, an increase in population pressure has resulted in land use changes and the migration of people into formerly untouched areas increasing the incidence of fire (Poffenberger and McGean, 1996). All areas, not only those with fire-prone vegetation types like tropical grasslands, savannas, and dry deciduous forests, have experienced decreases in tree cover and density.

Another hazardous side effect of the widespread burning practices in India is air pollution and the contribution of atmospheric aerosols and trace-gas emissions. Carbonaceous or black carbon (BC) aerosols cause strong atmospheric heating and large surface cooling that is as important to south Asian climate forcing as greenhouse gases (Babu et al., 2002; Badarinath et al., 2009b). The phenomenon is most persistent in south Asia, and has also been identified around large metropolitan areas in many tropical locations (Andreae et al., 2005). The continuous increase of BC aerosols over northern India and the southern slopes of the Himalayas has resulted in the formation the dense "brown clouds" that plague south Asia each winter. The clouds affect precipitation patterns, cause the melting of Himalayan glaciers, and result in increased heating rates in the lower and midtropospheres (Gautam et al., 2009; Gautam et al., 2010; Lau et al., 2006; Prasad et al., 2009; Ramanathan et al., 2001). The emissions from rural fire activities may have significant implications on atmospheric chemistry, climatic changes, and human health (Badarinath et al., 2009a; Vadrevu et al., 2011; Sharma et al., 2010). A thick blanket of haze formed by a mix of wood and dung fires, fossil fuel burning, wildfires, agricultural fires, and dust has been seen to develop during the beginning of winter from air masses building up against the Himalayas under continuous west wind situations. Lawrence and Lelieveld (2010) identified the largest concentrations of BC aerosols over the Indian subcontinent with the main sources being fossil fuel and biomass burning. Widespread burning of biomass, such as dried twigs, leaves, and dung and agricultural slash-and-burn practices, are common across poor, rural Asian areas where extensive forest and agricultural fires occur in India, especially in the north western part during

October and November each year due to crop residue burning (Sharma et al., 2010). Badarinath et al. (2009a) identified biomass-burning aerosols over Arunachal Pradesh during shifting cultivation practices. Gustafsson et al. (2009) suggested that biomass burning and particularly small-scale burning practices are the main cause of the dense brown clouds that affect the health of people inhaling the pollutants that are causing bronchitis and asthma. Smoke from agricultural burning contains numerous substances that can harm human health, including carbon monoxide, nitrogen oxides, and particulate matter (Lawrence and Lelieveld, 2010). The soot in the brown haze can be potentially linked to the hundreds of thousands of deaths each year mainly from lung and heart diseases (UNEP, 2011). Additionally, the brown cloud acts as a "global dimmer" through absorbing heat trapped by greenhouse gases. It also affects the regional climate, crop growth, and glacier melting through extended dry spells and increasing wildfire events (Ramanathan et al., 2007).

9.6 OUTLOOK

From our review, it is obvious that wildfires in India play a minor role as a hazard but are an essential tool for landscape management. However, the air pollution caused by agricultural burning not only impacts the health of people in close proximity to the burned fields but also those who live at great distances from them. To prevent such fires, we need to discover/develop alternative agricultural systems that do not require postharvest burning. An alternative system will likely mean more work for farmers and cost more to implement than the existing agricultural burning systems. Given this, it will be a challenge to convince farmers to change their practices.

It is most important to prevent agricultural fires from escaping into adjacent forests. This requires planning, basic fire knowledge, and strict regulation. For farmers who are cultivating encroached forestlands, the control of fire is certainly more difficult, and the risk of forest fires is greater compared to farmers burning agricultural land outside of the forest. But if wildfires are to be prevented, the cultivation of encroached forestlands should be banned.

Although accidental forest fires provide services from the forest, they are not necessary services; therefore, accidental forest fires can be seen as a less complex issue than fires put to a landscape by multiple stakeholders with a number of different motivations. When looking at the broad range of uses for fires, it appears to be impossible to totally avoid wildfires in India. There will always be somebody who requires wildfires for one reason or the other.

The aim of a national fire policy therefore cannot be to absolutely prevent wildfires, but rather to reduce their number and regulate the use of fire. The greatest chance of success for this policy lies within uncomplicated social settings, for example, where only one group of people burn the forest for one reason. The implementation of a fire reduction program in such places requires a number of key pieces of information upfront. First, how much land is required by the people to fulfill those needs that require the use of fire? Second, when fires are eliminated/reduced, there will be a response in the vegetation and we should be able to estimate/calculate future fuel loads and the susceptibility of the new ecosystem to fire. This knowledge will also allow for silvicultural operations toward demanded forest formations beyond fire-prone stages. Third, it is necessary that the forest functions and ecosystem services of the future forest are known to be able to plan their role in landscape management. Such an approach will allow for watershed or landscape-based planning that optimizes the management of the landscape in terms of its provision of ecosystem services.

A national or statewide fire prohibition policy will not achieve the desired results. Instead, a new fire policy must be worked out, one that provides a frame work allowing for many different fire applications based on the knowledge about local social settings, motivation for the use of fire, and the response of ecosystems to fire regimes.

It is important that decision makers at all levels are informed about the current extent of wildfires in India, the impact they have on ecosystem functioning, and the unavoidable trade-offs regarding the supply of ecosystem services under a changing fire scenario. Even though research of this sort is limited in India, there is sufficient knowledge available from international literature sources. With this information, fire policies at the state and national levels can be established. The policies should allow for fire regimes that are adapted to local settings, and the policies are expected to reduce the area burned.

REFERENCES

Andreae, M.O., Jones, C.J., Cox, P.M., 2005. Strong present-day aerosol cooling implies a hot future. Nature 435, 1187−1190.

Babu, S.S., Satheesh, S.K., Moorthy, K.K., 2002. Aerosol radiative forcing due to enhanced black carbon at an urban site in India. Geophys. Res. Lett. 29, 1880.

Badarinath, K.V.S., Latha, K.M., Kiran Chand, T.R., Gupta, P.K., 2009a. Impact of biomass burning on aerosol properties over tropical wet evergreen forests of Arunachal Pradesh, India. Atmos. Res. 91, 87−93.

Badarinath, K.V.S., Kumar Kharol, S., Rani Sharma, A., 2009b. Long-range transport of aerosols from agriculture crop residue burning in Indo-Gangetic Plains—a study using LIDAR, ground measurements and satellite data. J. Atmos. Sol. Terr. Phys. 71, 112−120.

Badgley, C., 1984. Pleistocene faunal succession in India. In: Whyte, R.O. (Ed.), Evolution of the East Asian Environment, vol. II. University of Hong Kong, Hong Kong, pp. 746−776.

Bahuguna, V.K., Upadhyay, A., 2002. Forest fires in India: policy initiatives for community participation. Int. For. Rev. 4, 122−127.

Blasco, F., 1983. The transformation from open forest to savanna in continental Southeast Asia. In: Bourlière, F. (Ed.), Ecosystems of the World (13, Tropical Savannas. Elsevier Scientific Publishing Company, Amsterdam, Oxford, New York, pp. 167−182.

Bond, W.J., van Wilgen, B.W., 1996. Fire and Plants, Population and Community Biology. Chapman & Hall, London.

Bor, N.L., 1938. The vegetation of the Nilgiris. Indian For. 64, 600–609.

Brandis, D., 1897/1994. Forestry in India, Origins and Early Developments. Natraj, Dehradun, India.

Carlson, P.C., Tanner, G.W., Wood, J.M., Humphrey, S.R., 1993. Fire in key deer habitat improves browse, prevents succession, and preserves endemic herbs. J. Wildl. Manage. 57, 914–928.

Chakraborty, M., 1994. Analysis of the causes of deforestation. In: Brown, K., Pearce, D.W. (Eds.), The Causes of Tropical Deforestation—The Economic and Statistical Analysis of Factors Giving Rise to the Loss of the Tropical Rain Forests. UCL Press, London, pp. 226–238.

Cochrane, M.A., 2001. Synergistic interactions between habitat fragmentation and fire in tropical forests. Conserv. Biol. 15, 1515–1521.

Cochrane, M.A., 2002. Spreading like Wildfire—Tropical Forest Fires in Latin America & the Caribbean: Prevention, Assessment and Early Warning. UNEP, Mexico City, Mexico.

Cole, M.M., 1986. The Savannas. Academic Press, London.

Favier, C., Chave, J., Fabing, A., Schwartz, D., Dubois, M.A., 2004. Modelling forest—Savanna mosaic dynamics in man-influenced environments: effects of fire, climate and soil heterogeneity. Ecol. Modell. 171, 85–102.

Forest Department Madras Presidency, 1923. Annual Administration Reports 1922–1923. Government Press, Madras, India.

Forest Survey of India (FSI), 1995. The State of the Forest Report. FSI, Ministry of Environment and Forests, Government of India, Dehra Dun, India.

Forest Survey of India (FSI), 2011. India State of the Forest Report 2011. FSI, Ministry of Environment and Forests, Government of India, Dehra Dun, India.

Furley, P.A., Rees, R.M., Ryan, C.M., Saiz, G., 2008. Savanna burning and the assessment of long-term fire experiments with particular reference to Zimbabwe. Prog. Phys. Geogr. 32, 611–634.

Gadgil, M., Meher-Homji, V.M., 1985. Land use and productive potential of Indian Savanna. In: Tothill, V.M., Mott, J.C. (Eds.), Ecology and Management of World's Savannas. Australian Academy of Science, Canberra, pp. 107–113.

Ganesan, R., Setty, R., 2004. Regeneration of amla, an important non-timber forest product from southern India. Conserv. Soc. 2, 365–375.

Gautam, R., Hsu, N.C., Lau, K.-M., 2010. Pre-monsoon aerosol characterization and radiative effects over the Indo-Gangetic Plains: implications for regional climate warming. J. Geophys. Res. 115, D17208. http://dx.doi.org/10.1029/2010JD013819.

Gautam, R., Hsu, N.C., Lau, K.-M., Tsay, S.-C., Kafatos, M., 2009. Enhanced pre-monsoon warning over the Himalayan-Gangetic region from 1979 to 2007. Geophys. Res. Lett. 36, L07704. http://dx.doi.org/10.1029/2009GL037641.

Goldammer, J.G., 1978. Feuerökologie und Feuer-Management-Freiburger Waldschutz Abh, 1, 2, Hrsg. v. Forstzool. Inst. d. Univ. Freiburg.

Goldammer, J.G., 1993. Feuer in Waldöksystemen der Tropen und Subtropen. Birkhäuser, Basel, Boston, Berlin.

Goldammer, J.G., Mueller-Dombois, D., 1990. Fire in tropical ecosystems and global environmental change: an introduction. In: Goldammer, J.G. (Ed.), Fire in the Tropical Biota. Ecological Studies, 84. Springer-Verlag Berlin, Heidelberg, pp. 1–11.

Gott, B., 2005. Aboriginal fire management in south-eastern Australia: aims and frequency. J. Biogeogr. 32, 1203–1208.

Government of India, 1999. In: National Forestry Action Programme—India, vol. 1. Ministry of Environment and Forests, New Delhi.

Gustafsson, Ö., Kruså, M., Zencak, Z., Sheesley, R.J., Granat, L., Engström, E., et al., 2009. Brown clouds over South Asia: biomass or fossil fuel combustion? Science 323, 495−498. http://dx.doi.org/10.1126/science.1164857.

Hawbaker, T.J., Radeloff, V.C., Syphard, A.D., Zhu, Z., Stewart, S.I., 2008. Detection rates of the MODIS active fire product in the United States. Remote Sens. Environ. 112, 2656−2664.

Hesmer, H.L., 1986. Einwirkung der Menschen auf die Wälder der Tropen. Westdeutscher Verlag, Opladen.

Hopkins, B., 1992. Ecological processes at the forest−Savanna boundary. In: Furley, P.A., Proctor, J., Ratter, J.A. (Eds.), Nature and Dynamics of Forest−Savanna Boundaries. Chapman & Hall, London, Glasgow, New York, Melbourne, Madras, pp. 21−62.

Hough, J., 1993. Why burn the bush? Social approaches to bush-fire management in West African national parks. Biol. Conserv. 65, 23−28.

Hunter, J.R., 1981. Tendu (*Diospyros melanoxylon*) leaves, bidi cigarettes, and resource management. Econ. Bot. 35, 450−459.

Johnsingh, A.J.T., 1986. Impact of fire on wildlife ecology in two dry deciduous forests in South India. Indian For. 112, 933−938.

Karanth, K.K., Curran, L.M., Reuning-Scherer, J.D., 2006. Village size and forest disturbance in Bhadra wildlife sanctuary, Western Ghats, India. Biol. Conserv. 128, 147−157.

Keeley, J.E., Bond, W.J., 1999. Mast flowering and semelparity in bamboos: the bamboo fire cycle hypothesis. Am. Nat. 154, 383−391.

Keeley, J.E., Bond, W.J., 2001. On incorporating fire into our thinking about natural ecosystems: a response to Saha and Howe. Am. Nat. 158, 665−670.

Keeley, J.E., Rundel, P.W., 2005. Fire and the Miocene expansion of C_4 grasslands. Ecol. Lett. 8, 683−690.

Keeley, J.E., Pausas, J.G., Rundel, P.W., Bond, W.J., Bradstock, R.A., 2011. Fire as an evolutionary pressure shaping plant traits. Trends Plant Sci. 16, 406−411.

Kepe, T., Scoones, I., 1999. Creating grasslands: social institutions and environmental change in Mkambati area, South Africa. Hum. Ecol. 27, 29−53.

Kharol, S.K., Badarinath, K.V.S., Sharma, A.R., Mahalakshmi, D.V., Singh, D., Prasad, V.K., 2012. Black carbon aerosol variations over Patiala city, Punjab, India—a study during agriculture crop residue burning period using ground measurements and satellite data. J. Atmos. Sol. Terr. Phys. 84−85, 45−51.

Kingwell-Banham, E., Fuller, D.Q., 2012. Shifting cultivators in South Asia: expansion, marginalisation and specialisation over the long term. Quat. Int. 249, 84−95.

Kohli, A., 2010. The Social and Political Context of Forest Fires: A Case Study in Andhra Pradesh, South India (Master's thesis). ETH Zurich.

Krishna, P.H., Reddy, C.S., 2012. Assessment of increasing threat of forest fires in Rajasthan, India using multi-temporal remote sensing data (2005−2010). Curr. Sci. (00113891) 102, 1288−1297.

Kurian, M., Singh, B., 1996. Role of Community Institutions in Fire Control in Haryana. Tata Energy Research Institute, New Delhi.

Lal, P., 2012. Estimating the size of tendu leaf and bidi trade using a simple back-of-the-envelope method. AMBIO 41, 315−318.

Landsberg, J.D., Lehmkuhl, J.F., 1995. Tigers, rhinos and fire management in India. First Conference on Fire Effects on Rare and Endangered Species and Habitats. IAWF 1997, Coeur d'Alene, Idaho.

Laris, P., 2002. Burning the seasonal mosaic: preventative burning strategies in the wooded Savanna of southern Mali. Hum. Ecol. 30, 155−186.

Lau, K.M., Kim, M.K., Kim, K.M., 2006. Asian summer monsoon anomalies induced by aerosol direct forcing: the role of the Tibetan Plateau. Clim. Dyn. 26, 855−864.

Lawrence, M.G., Lelieveld, J., 2010. Atmospheric pollutant outflow from southern Asia: a review. Atmos. Chem. Phys. 10, 11017−11096.

Lü, X.T., Lü, F.M., Zhou, L.S., Han, X., Han, X.G., 2012. Stoichiometric response of dominant grasses to fire and mowing in a semi-arid grassland. J. Arid Environ. 78, 154−160.

Main, M.B., Richardson, L.W., 2002. Response of wildlife to prescribed fire in Southwest Florida pine flatwoods. Wildl. Soc. Bull. 30, 213−221.

Mammen, E., 1964. Wirken deutscher Forstwirte in Übersee vor 1914. Forstarchiv 35, 117−123.

Mataix-Solera, J., Arcenegui, V., Tessler, N., Zornoza, R., Wittenberg, L., Martínez, C., et al., 2011. Soil properties as key factors controlling water repellency in fire-affected areas: evidences from burned sites in Spain and Israel. CATENA 108, 9−16.

Mathur, Y.K., 1984. Cenozoic palynofossils, vegetation, ecology, and climate of the north and north western sub-Himalayan region, India. In: Whyte, R.O. (Ed.), Evolution of the East Asian environment, vol. II. University of Hong Kong, Hong Kong, pp. 504−552.

Mbatha, K.R., Ward, D., 2010. The effects of grazing, fire, nitrogen and water availability on nutritional quality of grass in semi-arid Savanna, South Africa. J. Arid Environ. 74, 1294−1301.

Misra, R., 1983. Indian Savannas. In: Bourlière, F. (Ed.), Tropical Savannas. Ecosystems of the World, vol. 13. Elsevier Scientific, Amsterdam, Oxford, New York, pp. 151−166.

Mistry, J., Berardi, A., Andrade, V., Kraho, T., Kraho, P., Leonardos, O., 2005. Indigenous fire management in the cerrado of Brazil: the case of the Kraho of Tocantins. Hum. Ecol. 33, 365−386.

Moe, S.R., Wegge, P., 1997. The effects of cutting and burning on grass quality and axis deer (Axis axis) use of grassland in lowland Nepal. J. Trop. Ecol. 13, 279−292.

Myers, R.L., 2006. Living with Fire—Sustaining Ecosystems and Livelihoods through Integrated Fire Management. The Nature Conservancy's Global Fire Initiative, Arlington, VA.

Nanda, P.K., Sutar, P.C., 2003. Management of Forest Fire through Local Communities: A Study in the Bolangir, Deogarh and Sundergarh Districts of Orissa, India. Community-Based Fire Management: Case Studies from China, the Gambia, Honduras, India, the Lao People's Democratic Republic and Turkey. Food and Agriculture Organization of the United Nations Regional Office for Asia and the Pacific, Bangkok, Thailand.

Noble, W.A., 1967. The shifting balance of grasslands, shola forests, and planted trees in the upper Nilgiris, southern India. Indian For. 93, 691−693.

Poffenberger, M., McGean, B., 1996. In: Village Voices, Forest Choices—Joint Forest Management in India. Oxford University Press, Oxford, UK.

Prasad, A.K., Yang, K.-H.S., El-Askary, H.M., Kafatos, M., 2009. Melting of major glaciers in the western Himalayas: evidence of climatic changes from long term MSU derived tropospheric temperature trend (1979−2008). Ann. Geophys. 27, 4505−4519.

Puyravaud, J.P., Shridhar, D., Gaulier, A., Aravajy, S., Ramalingam, S., 1995. Impact of fire on a dry deciduous forest in the Bandipur National Park, southern India: preliminary assessment and implications for management. Curr. Sci. 68, 745−751.

Pyne, S.J., 1994. Nataraja: India's cycle of fire. Environ. Hist. Rev. 18, 1−20.

Pyne, S.J., 1995. Cycle of Fire. University of Washington Press and Nataraja (India), Seattle, London, Dehra Dun (India).

Pyne, S.J., 1997. In: Vestal Fire: An Environmental History, Told through Fire, of Europe and Europe's Encounter with the World, vol. 659. University of Washington Press, Seattle.

Rakyutitham, A., 2000. Forest fire: a history of repression and resistance. Watershed 6 (2), 35−42.

Ramakrishnan, P.S., 2007. Traditional forest knowledge and sustainable forestry: a north-east India perspective. For. Ecol. Manage. 249, 91−99.

Ramanathan, V., Crutzen, P.J., Lelieveld, J., Mitra, A.P., Althausen, D., Anderson, J., et al., 2001. Indian Ocean experiment: an integrated analysis of the climate and the great Indo-Asian haze. J. Geophys. Res. 106, 28371−28398.

Ramanathan, V., Ramana, M.V., Roberts, G., Kim, D., Corriganm, C., Chung, C., et al., 2007. Warming trends in Asia amplified by brown cloud solar absorption. Nature 448, 575−578.

Ratnam, J., Bond, W.J., Fensham, R.J., Hoffmann, W.A., Archibald, S., Lehmann, C.E.R., et al., 2011. When is a 'forest' a Savanna, and why does it matter? Global Ecol. Biogeogr., 1−8. http://dx.doi.org/10.1111/j.1466-8238.2010.00634.x.

Rist, L., Uma Shaanker, R., Milner-Gulland, E.J., Ghazoul, J., 2010. The use of traditional ecological knowledge in forest management: an example from India. Ecol. Soc. 15, 3.

Roveta, R., 2008. Traditional Use of Fire for the Provision of Ecosystem Services: A Case Study in BRT Wildlife Sanctuary (Master's thesis). University of Freiburg/Br.

Saha, S., 2002. Anthropogenic fire regime in a deciduous forest of central India. Curr. Sci. 82, 101−104.

Saha, S., 2003. Patterns in woody species diversity, richness and partitioning of diversity in forest communities of tropical deciduous forest biome. Ecography 26, 80−86.

Saha, S., Howe, H.F., 2001. The bamboo fire cycle hypothesis: a comment. Am. Nat. 158, 659−663.

Saha, S., Howe, H.F., 2003. Species composition and fire in a dry deciduous forest. Ecology 84, 3118−3123.

Saigal, R., 1990. Modern forest fire control: the Indian experience. Unasylva 162, 21−27.

Schmerbeck, J., 2003. Patterns of Forest Use and its Influence on Degraded Dry Forests: A Case Study in Tamil Nadu, South India. Shaker Verlag, Aachen.

Schmerbeck, J., Seeland, K., 2007. Fire supported forest utilisation of a degraded dry forest as a means of sustainable local forest management in Tamil Nadu/South India. Land Use Policy 24, 62−71.

Schmerbeck, J., Kohli, A., Seeland, K., in press. Ecosystem services and forest fires in India - Context and Policy implications from a Case Study in Andhra Pradesh, Forest Policy and Economics.

Semwal, R.L., Chatterjee, S., Punetha, J.C., Pradhan, S., Dutta, P., Soni, S., et al., 2003. Forest Fires in India—Lessons from Case Studies. World Wildlife Fund for Nature-India, India.

Shaffer, L.J., 2010. Indigenous fire use to manage Savanna landscapes in southern Mozambique. Fire Ecol. 6, 43−59.

Shakesby, R.A., 2011. Post-wildfire soil erosion in the Mediterranean: review and future research directions. Earth Sci. Rev. 105, 71−100.

Shakesby, R.A., Wallbrink, P.J., Doerr, S.H., English, P.M., Chafer, C.J., Humphreys, G.S., et al., 2007. Distinctiveness of wildfire effects on soil erosion in south-east Australian eucalypt forests assessed in a global context. For. Ecol. Manage. 238, 347−364.

Sharma, A.R., Kharol, S.K., Badarinath, K.V.S., Singh, D., 2010. Impact of agriculture crop residue burning on atmospheric aerosol loading—a study over Punjab State, India. Ann. Geophys. 28, 367−379.

Singh, C., Panigrahy, S., 2011. Characterisation of residue burning from agricultural system in India using space based observations. J. Indian Soc. Remote Sens. 39, 423−429.

Sinha, A., Brault, S., 2005. Assessing sustainability of non-timber forest product extractions: how fire affects sustainability. Biodiversity and Conserv. 14, 3537−3563.

Springate-Baginski, O., Blaikie, P., 2007. In: Forest, People and Power—The Political Ecology of Reform in South Asia. Earthscan, London, UK.

Stebbing, E.P., 1922. In: The Forests of India, vol. I. John Lane the Bodley Head, London.

Takahata, C., Amin, R., Sarma, P., Banerjee, G., Oliver, W., Fa, J., 2010. Remotely-sensed active fire data for protected area management: eight-year patterns in the Manas National Park, India. Environ. Manage. 45, 414—423.

Thomas, S.M., Palmer, M.W., 2007. The montane grasslands of the Western Ghats, India: community ecology and conservation. Community Ecol. 8, 67—73.

Tomor, B.M., Owen-Smith, N., 2002. Comparative use of grass regrowth following burns by four ungulate species in the Nylsvley Nature Reserve, South Africa. Afr. J. Ecol. 40, 201—204.

UNEP, 2011. Integrated Assessment of Black Carbon and Tropospheric Ozone: Summary for Decision Makers. United Nations Environment Programme & World Meteorological Association, Nairobi, Kenya.

Vadrevu, K., Badarinath, K.V.S., Anuradha, E., 2008. Spatial patterns in vegetation fires in the Indian region. Environ. Monit. Assess. 147, 1—13.

Vadrevu, K.P., Ellicott, E., Badarinath, K.V.S., Vermote, E., 2011. MODIS derived fire characteristics and aerosol optical depth variations during the agricultural residue burning season, north India. Environ. Pollut. 159, 1560—1569.

Vadrevu, K.P., Csiszar, I., Ellicott, E., Giglio, L., Badarinath, K.V.S., Vermote, E., et al., 2013. Hotspot analysis of vegetation fires and intensity in the Indian region. IEEE J. Sel. Top. Appl. Earth Obs. Remote Sens. 6, 224—238.

Vayda, A.P., 1996. Methods and Explanations in the Study of Human Actions and Their Environmental Effects. Center for International Forestry Research and World Wide Fund for Nature, Jakarta.

Veldman, J.W., Mattingly, W.B., Brudvig, L.A., 2012. Understory plant communities and the functional distinction between Savanna trees, forest trees, and pines. Ecology 94, 424—434.

Wickens, G.E., 1994. Sustainable management for non-wood forest products in the Tropics and Subtropics. In: Readings in Sustainable Forest Management, FAO Forestry Paper, vol. 122. FAO, Rome, pp. 55—65.

Wisner, B., Blaikie, P., Cannon, T., Davis, I., 2004. At Risk—Natural Hazards, People's Vulnerability and Disasters. Routledge, Wiltshire.

World Bank, 2006. India, Unlocking Opportunities for Forest-Dependent People in India. Main report. South Asia Region Report No. 34481-IN. 1.

Yadav, M., Dugaya, D., 2013. Non-timber forest products certification in India: opportunities and challenges. Environ. Dev. Sustainability 15, 567—586.

Chapter 10

System of Wildfires Monitoring in Russia

Evgeni I. Ponomarev

V.N. Sukachev Institute of Forest, Siberian Branch of Russian Academy of Sciences, Akademgorodok, Krasnoyarsk, Russia; Siberian Federal University, pr. Svobodnyi, Krasnoyarsk, Russia

Valeri Ivanov

Siberian State Technological University, Krasnoyarsk, Russia

Nikolay Korshunov

Russian Institute of Continuous Education in Forestry, Pushkino, Moscow obl., Russia

ABSTRACT

Wildfire is one of the main natural disturbances in Russian forests, with some 2−17 million hectares being burned annually. The trend of large-scale wildfires, with areas burnt exceeding 2,000 ha, having increased over recent years, is making significant contributions to atmospheric emissions that contribute to climate change, and highlights the need for and complexity of fire risk management and monitoring. The challenges faced in monitoring and risk management can be understood in relation to the natural conditions of forest environments in Russia. Given the vast nature of the geographical area to be managed, satellite techniques are the primary means for wildfire monitoring in most part of the boreal forest zone of Russia. This increases both the efficiency of wildfire detection and the capacity to obtain information on wildfire attributes. The chapter then proceeds to discuss how the scale and remoteness of much wildfire activity, a lack of transport routes, and low density of population in northern regions of Siberia results in a high level of burn and a significant number of extreme large-scale wildfires annually. The chapter also discusses how fire risk management must accommodate the modern state politics in forestry, climate change, and the growing anthropogenic impact on fire risk.

Wildfire Hazards, Risks, and Disasters. http://dx.doi.org/10.1016/B978-0-12-410434-1.00010-5
187

10.1 INTRODUCTION

One of the main natural disturbances in Russian forests is wildfires. The annual burned area makes up 2–17 million hectares (Conard et al., 2002; Soja et al., 2004). The importance of managing this hazard derives from the observed increase in wildfire activity and risk, and the consequent increasing in the area burned that has been observed during previous years (Loupian et al., 2006; Sukhinin, 2008; Shvidenko et al., 2011; Goldammer, 2013). For example, in 2010, extreme wildfires in the European part of Russia resulted in 2.2 million hectares of forests being damaged (Bondur, 2011). At the same time, there was more than 5,000 wildfires in the forest zone and over 10,000 wildfires in the steep zone of Eastern Siberia resulting in a total of about 8 million hectares of forest being burned (Bartalev et al., 2012; Ponomarev and Shvetsov, 2013). Overall, there were in excess of 24,000 wildfires detected in 2011 and 23,000 wildfires detected in 2012, including 2,200 large-scale fires in boreal forests. As a result, wildfires produced extreme emissions to the atmosphere in 2012 in Siberia. The concentration of CO_2 was higher, up to 35 times of the average according to measurements at the Zotino Tall Tower Observatory (ZOTTO) (Panov et al., 2012).

The trend of large-scale wildfires, with areas burnt exceeding 2,000 ha, has increased over recent years. Large-scale wildfires have occurred mainly in the territory of northern boreal forests of the Ural, of Siberia, and of the Far Eastern Federal Districts of Russia (see Figure 10.1). In 2012 and 2013, most large-scale wildfires in Russia have occurred in Yakutia, Evenkiya, Khanty-Mansyisk, and north of Krasnoyarsk regions (Rubtsov et al., 2011, Ponomarev and Shvetsov, 2013). This escalating pattern of risk has increased the need for risk management and monitoring.

Given the vast nature of the geographical area to be managed, satellite techniques are the primary means for wildfire monitoring in most parts of the boreal forest zone of Russia. Along with on-ground and air monitoring, the use of satellite methods and products facilitate the expansion of the area capable of being monitored. This increases both the efficiency of wildfire detection and the capacity to obtain information on wildfire attributes (e.g., coordinates estimation of active burning area, total damaged areas, temporal characteristics of fires). This work is especially important for forests in the northern territories of Siberia. These forests in the boreal zone have significant economic and ecological importance as one of the major components of the global carbon budget (Sukhinin et al., 2003; Neigh et al., 2013).

In the current environments the fire fighting system is not actually used in the boreal forests of Siberia. This zone is mostly unpopulated. On-ground monitoring as far as aircraft observing is used in practice over the European part of Russia and southern territories of Siberia and Russian Far East. The lack of transport routes as well as the low population density in northern regions of Siberia results in a high level of burning and a significant number of extreme large-scale wildfires annually.

FIGURE 10.1   Wildfires detected by satellite technique in 2010–2013. Federal districts of Russia: 1, the Central; 2, the Northwestern; 3, the Southern and the Northern Caucasus; 4, the Trans-Volga; 5, the Ural; 6, the Siberian; 7, the Far Eastern.

In the current chapter we will discuss the current wildfire situation in Russia. It is in close connection with modern state politics in forestry and climate changing, as well as under strong influence of anthropogenic impact. The challenges faced in monitoring and risk management can be understood in relation to the natural conditions of forest environments in Russia.

10.2 NATURAL CONDITIONS AND FORESTS IN RUSSIA

According to the Federal Forestry Agency of the Russian Federation (http://www.rosleshoz.gov.ru) between 1 and 80 percent of different areas of Russia is covered by forests (see Figure 10.2). The mean percentage of forested area in Russia is 46.6 percent. The highest level of forest cover is in the Siberian Federal District (53.9%) followed by the Northwest Federal District (52.2%), and the Far East Federal District (48%). The Central, the Trans-Volga, and the Ural Federal Districts have about 36% of forest-covered area, and the Southern and the Northern Caucasus Federal Districts have 6.2 percent and 10.9 percent forest-covered area, respectively.

Forest-covered land in Russia is characterized by a wide range of forest-growing conditions associated with latitudinal zonation and longitudinal provincialism of climate and soils. Additionally, there is variation of forest vegetation associated with the vertical zonation of climate and soil in

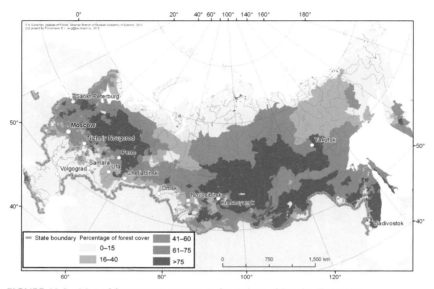

FIGURE 10.2 Map of forest cover percentage for regions of Russian Federation.

FIGURE 10.3 Schematic map of natural zones on the territory of Russia. Numbers of natural zones are: 1, nonforested area; 2, pretundra forests and sparse taiga; 3, taiga; 4, area of coniferous and deciduous forests; 5, forest steppe zone; 6, steppe; 7, semidesert and desert; 8, mountainous area of the northern Caucasus; 9, south Siberian mountain area.

the mountainous regions of Russia (which occupies about a quarter of the territory).

Steppe and forest steppe zone extends along the southern border of Russia (see Figure 10.3). Significant amounts of nonforest burning are observed here annually as steppe and agricultural fires (see Figure 10.1 and Figure 10.2). The vegetation cover of the forest zone of Russia includes taiga and mixed coniferous broad-leaved forests (the latter predominantly within the southern regions of the European and Far Eastern parts of Russia). Boreal forests in north–south direction are divided into northern, middle, and southern taiga. From west to east, taiga forests can be classified according to the level of wildfire impact. Thus, wildfires occur in the taiga of the European part of Russia once every 10–30 years (Bondur, 2011) while taiga forests of Siberia are under the strong influence of large-scale wildfires annually. The level of wildfire hazard is high in the taiga of the Far East, including Sakhalin and Kamchatka.

Within these zonal–provincial units, dark coniferous (pine) and light coniferous (larch) forests are distributed in accordance with the characteristics of their dominant species. Dark coniferous stands predominate in the taiga zone of the European part of Russia, Western Siberia, and in the mountains of southern Siberia, including the Yenisei River Ridge and the Lena River Plateau. Pine forests predominate in the taiga zone of the European part of Russia; Western, Central, and Eastern Siberia; in the West Siberian forest steppe areas; in the lowlands of Trans-Baikal region; and along Amur River. Larch forests are common in the northern taiga of Western Siberia, the taiga forests of Central and Eastern Siberia, and the Far East, including the mountains of southern Siberia.

Geospatial analysis of wildfire occurrence has been produced for forests predominated by different tree stands. During 2010–2012 up to 26 percent of total burned area was fire scars in larch stands. This is largely determined by a widespread occurrence of larch in Siberia (Kharuk et al., 2013). Furthermore, a large number of wildfires occurred in pine forests in areas with high population density. Between 7–13 percent of the total areas damaged by wildfires annually are pine forests. Spring and autumn grass fires in deciduous forests contribute a further 7–10 percent to the overall statistics of burning areas. And wildfires in dark conifers stands contribute from 3.5 percent up to 7 percent.

The construction of the Russian system of forest fire protection needs to be developed in accordance with the prevailing natural conditions and forest distribution. The main challenges of fire preventing organizing are: wide distributing of fighting forces and resources; zoning of forest fire monitoring type; differentiation of tactics, strategy, and technology of fire extinguishing; failing of system for early fire detection within wide territories; and low density of road network in most parts of boreal forests. The risk management challenges that arise from the aforementioned distribution patterns become evident when reviewing fire distribution.

10.3 FIRE HISTORY AND CURRENT STATISTICS

According to long-term official data (http://www.rosleshoz.gov.ru/forest_fires/), more than 450,000 wildfires occurred in forests of the Russian Federation from 1969 to 2007. On an average there were over 12,000 fire events during each fire season. After the collapse of the USSR since 1993 till 2007 there are more than 350,000 fires (Federal Agency of Forestry, 2008). The number of fires ranged from 16,000 in 2007 to 37,000 in 2002. Between 1993 and 2007, the burned area exceeded 15 million hectares, on an average about 1 million hectares per year. The maximum burned area from fires was observed in 1998 with more than 2.2 million hectares burned. The minimum was in 1995 with about 0.35 million ha burned (see Figure 10.4). According to official statistics (Federal Agency of Forestry, 2008), the average area burnt per fire event was 45 ha per year in the Russian Federation, but this varies from 14 ha up to 94 ha.

The average number of wildfires in large areas has increased in the past decades (Shvidenko and Shchepashchenko, 2013). Especially in forests of the boreal zone of Siberia and Russian Far East the increase is most likely due to a reduction of the effectiveness of the prophylactic and preventive measures under current conditions of lower funding (Isaev and Korovin, 2013; Goldammer, 2013). Currently, this area is only monitored by satellites. Because there are no active air bases, forest observation by aircraft is irregular. As a result, early detection of fires is impossible. Further, wildfire fighting is carried out only in cases of threats to human settlements. However, it should also be noted that the number of fires is increasing due to better accessibility of forests for people (e.g., increases in the number of vehicles, increases in the density of roads in at-risk areas), the use of forests for recreation, and harvesting of forests in remote and inaccessible areas (Federal Agency of Forestry, 2008). Two latter issues attain additional significance as a result of their influences on the cause of fires.

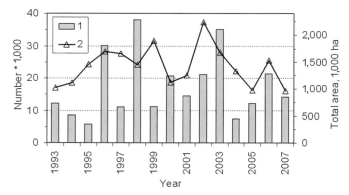

FIGURE 10.4   The dynamics of forest fires in the Russian Federation for the period 1993–2007 (according to data of the Federal Forestry Agency). 1, The total burned area per year; 2, the number of wildfires.

TABLE 10.1 The Distribution of Causes of Forest Fires
in Russia (1969–2007)

Fire Cause	Percentage of Total
Local population	55.3
Unknown	19.4
Thunderstorms	15.3
Agricultural burnings	5.9
Harvesting	1.8
Railroads	1.2
Field works of expeditions	0.6
Burning of woody debris	0.5

With regard to the causes of fire, forest fire statistics enable identification of several main causes (Kuzmichev, 2006; Andreev & Brukhanov, 2011). As the analysis of the data in Table 10.1 shows, some 53 percent of fires are directly or indirectly caused by local population activities. Remarkably, for a significant percentage of fires the cause is unknown—more than 19 percent. On an average, in the forests of Russia, some 15 percent of fires are caused by thunderstorms. In the sparsely populated areas of boreal taiga the percentage of fires caused by thunderstorms can be much higher—up to 70 percent of the total number of detected fires (Ivanov and Ivanova, 2010).

Summary information shown above allows representing the general dynamics of forest fires in Russia. Differences in the pattern of forest fire activity in the Russian Federation are determined by forest vegetative, climatic, economic, and, consequently pyrological conditions. The most expeditious way of examining this is through comparing regular differences in the distribution of forest fires by Federal Districts (see Figure 10.5). This analysis was performed using data for the period 1993–2007 (Federal Agency of Forestry, 2008). Most wildfires occur in the forest of the Siberian Federal District every year with, on an average, more than 30 percent of the total number of fires in the Russian Federation. In 2003, 50 percent of Russian wildfires happened in the Siberian Federal District. The second group includes the Ural, Central, and Northwestern Federal Districts, which accounted for 20 percent of the number of fires, but in some years this number reached extremely high values—up to 30 percent of fires in the forests of the Central Federal District in 2002 and more than 50 percent in the Ural Federal District in 2004. The third group in terms of the number of fires (10–20% of total) consists of the Trans-Volga, the Southern and Northern Caucasus, and the Far Eastern Federal Districts.

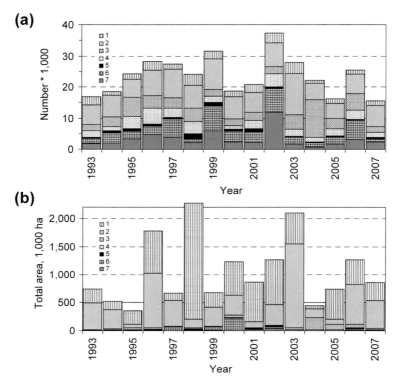

FIGURE 10.5 A comparison of the dynamics of the number of forest fires (a) and total burned areas (b) for federal districts of Russia. Federal districts are: 1, the Far Eastern; 2, the Siberian; 3, the Ural; 4, the Trans-Volga; 5, the Southern and the Northern Caucasus, 6, the Northwestern; 7, the Central.

Distribution, and thus the classification of the area, changes considerably when considering the total burned area as well as the average area burned per fire. For the period under consideration, more than 90 percent of the burned area falls in two federal districts: the Siberian and Far Eastern Districts (see Figure 10.5b). At the same time, in terms of the average burned area, the Far Eastern Federal District is the principle contributor to this distribution.

Since satellite monitoring of wildfires was started, alternative data is available (Sukhinin et al., 2003; Loupian et al., 2006; Kukavskaya et al., 2013; Ponomarev and Shvetsov, 2013). This information complements the data of official statistics and allows evaluating the dynamics of the fire situation, including the recent years (see Figure 10.6). Results of preprocessing of the satellite database of active fire detections are presented for 1996–2013. The database of wildfires over the Asian part of Russia is available in the format of GIS-layers (geographic information system layers) derived from NOAA/AVHRR (National Oceanic and Atmospheric Administration'

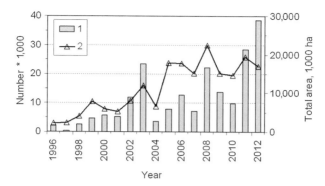

FIGURE 10.6   Statistics of satellite-detected wildfires (including forest and nonforest burning) for 1996–2012. 1, The aggregated area of burning polygons per year; 2, the number of detected wildfires.

Satellites/Advanced Very High Resolution Radiometer) and Terra/MODIS (NASA's Terra satellite/Moderate Resolution Imaging Spectroradiometer) imagery. According to long-term statistical data from satellite monitoring, more than 70 percent of total wildfires are due to nonforest burning in forest steppe and steppe zones. But up to 70 percent of annual burned area is the result of wildfire activity in forests.

Analysis of wildfire characteristics for the Asian part of Russia allows selecting two types of active fire observations: short-living wildfires and large-scale wildfires. Short-living fires are defined as those that are visible on only one satellite pass. Large-scale fires cover over 1,500–2,000 ha and may last for several weeks.

The average annual percentage of short-living fires is about 45–65 percent of the total number of fires and the area burned was up to 30 percent of the total annual burned area. As an example, during 2012 more than 50 percent of wildfires were recorded as 1-day agricultural burns. So their total area was initially estimated as up to 15 percent of the total fire scar area per year. This type of hot spot was detected primarily during spring (March to end of May). The spatial distribution of short-living fires was verified using SPOT (Satellite Pour l'Observation de la Terre) and Landsat imagery.

Statistics about large-scale wildfires allows defining increases in the number of fire events over larger areas annually. Mass extreme wildfires have occurred every 3–5 years in different regions of the Asian part of Russia during the past few decades. Annually, up to 90 percent of the total burned area is the result of large-scale wildfires, with about 45 percent of wildfires being in 100–200 ha range and up to 50 percent of wildfires being under 1,000 ha. Only 5 percent of wildfires are large-scale events in the boreal zone of Siberia (whose area can be over 2,000 ha) (Ponomarev et al., 2012; Ponomarev and Shvetsov, 2013).

Wildfires in Russia cause damage to forestry from $100 million to $400 million annually (Vorobjova et al., 2004; Binenko and Donchenko,

2004). Assessment in the previous years has shown that damage to forest is from $400 million in 2010 (Isaev and Korovin, 2013) up to $3000 million in 2012 and to $700 million in 2013 (Rosleskhoz, 2013).

The most catastrophic fire situation was reported in 2010 in the European part of Russia (Bondur, 2011; Isaev and Korovin, 2013). These fires affected 199 settlements. About 3,200 houses were completely destroyed, and more than 7,000 people lost their homes. Wildfires killed 62 people. Moreover, tens of thousands of people in Moscow died were of smog influence (patients with respiratory problems) (Smirnov, 2013). In total, in the European part of Russia 60 federal reserves and national parks suffered.

The extreme fire seasons of 2012 and 2013 (Ponomarev and Shvetsov, 2013) were characterized by high level of damage but did not result in loss of life or houses because the main part of the burned areas was in the non populated territories of Siberia. Statistically, Siberia and the Far East region faced high wildfire risk during the 1970s—1980s and during the first two decades of the 21st Century (Shvidenko and Shchepashchenko, 2013). Given the extensive area that needs to be managed, responsibility for risk management is a federal responsibility.

10.4 FEDERAL INSTITUTIONS OF WILDFIRE PREVENTING AND FIGHTING

Implementation of wildfire safety measures in forests and forest fire fighting is carried out in accordance with the Forest Code of the Russian Federation by state departments and organizations. The organizational structure of forest fire protection in Russia was modernized in 2012 (see Figure 10.7). Authority for forestry and forest fire fighting is entrusted to agencies in each region of the Russian Federation. All forests are federal property. They are divided into territorial forest districts, named Forestry ("Lesnichestvo" in Russian).

The Ministry of Emergency of the Russian Federation determines the general policy for forest fire safety, and is responsible for the operation of the state system for prevention and management of emergency situations, including hazards from wildfires near settlements, objects of infrastructures, and populated places. The Ministry of Natural Resources and Environment is responsible for organizing and managing the prevention of fires in forest territories, as well as with managing pest control and diseases of forest vegetation. The Federal Forestry Agency ("Rosleskhoz" in Russian) is the structural division of the Ministry of Natural Resources and Environment of the Russian Federation. It determines the policy and strategy for forest protection and controls the implementation of the actions of authorities, which have been delegated to subjects of the Russian Federation.

Annual earmarked interventions are funded by the federal budget to regions of Russia to facilitate the performance of authorities in forest areas in preventing and fighting wildfire. According to The Forest Code every

FIGURE 10.7   Organization structure of forest fire protection in Russia.

constituent entity of the Russian Federation should create and contain state agencies or should attract commercial organizations for forest fire fighting. Thus, regional forestry government organizations are involved in monitoring, preventing, and extinguishing wildfire in the forests. The Forest Code requires every region to develop an annual contingency plan to deal with forest fires. Forestry's fire fighting organizations are represented mainly by two types: forest fire centers, dominated by on-ground units (fire chemical stations), and airbases dominated by air units (airbase divisions).

The task of extinguishing wildfires is carried out by the different firefighter groups under the jurisdiction of the Russian Ministry of Emergencies outside the boundaries of forest lands and on the land of other categories, including settlements and urban areas. Municipal departments and volunteer fire departments are the dominant agency in country areas. However, Federal specialized units of "Rosleskhoz" and of the Russian Ministry of Emergencies could be used in the case of emergencies caused by forest fires. This complex management task is assisted by territory zoning and monitoring strategies.

10.5 TERRITORY ZONING ACCORDING TO THE TYPE OF FIRE MONITORING

The major peculiarities of Russia are the large distances and vast territory over which fire risk is spread, and the sparse population and poor infrastructure in these areas. Geographical factors have dominated the construction of the Russian system of forest fire protection. Therefore, in this system there is a clear separation of the protected area into monitoring zones and areas of forces and resource application. The establishment of areas for the application of fire fighting forces and resources determines the tactics, strategy, and technology of fire extinguishing, the organizational structure for the given area, types of extinguishing teams, the order of interaction, and leadership of fighting forces. There are 4 zones of forest fire monitoring in Russia today (Andreev & Brukhanov, 2011) (see Figure 10.8):

1. The zone of ground-based monitoring (about 7% of territory) is where the detection of forest fires is performed by ground-based methods: patrol, observation from uplands, the use of observation towers, and television masts.
2. The zone of air monitoring (up to 42% of forests) involves regular patrol by aircrafts and helicopters.
3. The zone of satellite monitoring of level 1 (almost 20% of forest lands) involves detection being performed by satellites and where forest fire extinguishing is performed selectively.
4. The zone of satellite monitoring of level 2 (about 31% of forest lands) entails detection being carried out by satellites, but where forest fire extinguishing is not performed, except in cases of settlement threats. Here, the area of forest fires is recorded, but is not displayed in the official statistics.

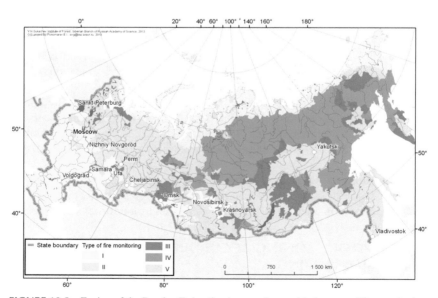

FIGURE 10.8   Zoning of the Russian Federation in accordance with the type of fire monitoring. Zones are: I, of ground-based monitoring; II, of air monitoring; III, of satellite monitoring of level 1, IV, of satellite monitoring of level 2; V, territories of protected areas (state reservations, national parks, etc.).

Monitoring strategies are accompanied by two kinds of fire fighting forces and resources application. The area of application of ground forces and resources for protecting forests focuses on densely populated areas with well-developed road network in the forests, where arrival of ground fighting forces within 3 h of the discovery of the forest fire is possible. The category of "large" forest fire includes fires having size of 25 ha or more (Andreev & Brukhanov, 2011).

Work in the areas of application of ground forces is carried out by fire chemical stations units, where there are trained teams equipped with off-road vehicles, fire engines, forest fire tractors, bulldozers, and fire boats. In 1,600 fire chemical stations there are about 50,000 firefighters and more than 10,000 units of firefighting equipment. Areas of ground-based monitoring and areas of application of ground forces and resources cover about 98 million hectares.

The application of air forces and means for forest protection focuses on remote areas with poor (or no) road network, where the arrival of ground forces and means of fire suppression within 3 h of the discovery of the forest fire is not possible. The category of "large" forest fire includes fires spreading over 200 ha.

The deployment of air forces and resources is preferably carried out by air units where there are teams of parachutist firefighters equipped with fire extinguishers, special machinery and equipment, and air transportation means. In 250 air divisions there are more than 5,000 aircraft firefighters. Annually up to 400 aircrafts sourced from commercial airlines are used. Annual regulatory requirement of aviation operations is at the level of 80,000−100,000 flight hours. Zone of air monitoring and the areas of aviation capabilities application cover about 493 million hectares. The zones of space monitoring include remote and hard-to-reach areas. Burned area of more than 2,000 ha is the criteria for classifying wildfires over the territory of satellite observation as large-scale burning (Andreev & Brukhanov, 2011).

All information about wildfires from on-ground, air, or satellite observation are collected in the data bank of the information system of remote monitoring ("ISDM-Rosleskhoz" in Russian) of the Federal Forestry Agency. This system was installed in 2005. The functioning of the information system for remote monitoring provides information to the federal agency of Aerial Forest protection ("Avialesookhrana" in Russian) (Pushkino, Moscow region, http://www.aviales.ru/). Any forestry organization can get free access to the data of wildfire monitoring system via Internet. More than 3,000 organizations, including private businessman who are renting a forest, use the services of this system.

Data from the satellite monitoring of wildfires is available for the past 20 years. The first center for wildfire monitoring was established in Krasnoyarsk in V.N. Sukachev Institute of Forest, Siberian Branch of Russian Academy of Sciences, in 1994−1995. Since then, alternative information on wildfire statistics is available also for territory of the Asian part of Russia. At the present time, a number of satellite receiving centers allow collecting data on satellite imagery and wildfire information from the whole territory of Russia. The information for the Federal Forestry Agency is collected in Pushkino (Moscow region), Novosibirsk, Krasnoyarsk, and Khabarovsk (Bartalev et al., 2012), while centers in Vologda, Krasnoyarsk and Vladivostok are used by the Russian Ministry of Emergencies since early 2000 (Kudrin and Resnikov, 2006). The importance of these resources is increasing as a result of growing fire risk.

10.6 NATURAL FIRE DANGER AND ANTHROPOGENIC IMPACT

The degree of natural fire risk, as well as the level of risk from anthropogenic sources and impact on forests, has been estimated using available information on physical geography, forestry, pyrological factors, and social factors of the territory (Ponomarev et al., 2010). These characteristics were accompanied with quantitative values so that the scale of the obtained values reflects differences for every Forestry of Russia.

Statistical analysis took into account the data on long-term forest fires monitoring from 1969 to 2007. It revealed the degree of connection between factors taken into account and fire regimes employed in the regions. This degree of connection has been considered using calibrating by weight coefficients. The value of factors, which is characterizing the level of anthropogenic impact and natural fire danger of forests, have been normalized per 100,000 hectares of the territory. The characteristics data were included in vector GIS layer as new attributive information for every Forestry territory. Thus characteristics of natural fire danger and anthropogenic influence potential are available for the whole of Russia in the scale of Forestry (see Figure 10.9).

The following factors have been used for the natural fire danger evaluation: data on categories of lands and land usage within the boundaries of the Forestry; data on predominated tree stands and forest type in connection to pyrological characteristics; the coefficient of thermal and moisture supply of territory, calculated on the basis of long-term temperature and precipitation data; and the percentage of lightning/human-ignited wildfires.

The following factors have been used for the level of anthropogenic impact evaluating: data on forest cover percentage; logging disturbance estimation; the density of transport network (including roads, forest roads, rivers); and estimation of recreational pressure on forest, population density, and distribution of settlements. This data was normalized per area of Forestry and was interpolated within the boundary of every Forestry.

The analysis revealed that the majority of the Forestry of the Central, the Southern, and the Northern Caucasus and the Trans-Volga Federal Districts are characterized by low to moderate levels of natural forest fire danger. Moderate

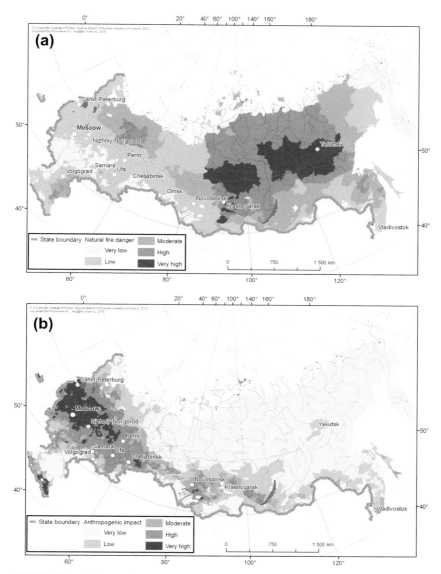

FIGURE 10.9   Scheme of the zoning of forest territory of Russia on the basis of natural fire danger (a) and anthropogenic impact (b) levels.

to high level of natural fire danger is corresponded to the most of the forest districts of the Siberian, the Northwestern, and the Far Eastern Federal Districts (see Table 10.2). In most cases these Forestries are under low anthropogenic influence. But the level of lightning fire danger is one of the main factors determining the forest fire occurrence there.

TABLE 10.2 Distribution of Forestries according to Natural Fire Danger
(Percentages of Total Number of Forestries in Federal District)

Natural Fire Danger Class	Level of Natural Fire Danger	North Western	Central	Trans-Volga	Southern and Northern Caucasus	Ural	Siberian	Far Eastern
					Federal District			
1	Very low	1	37	37	10	13	1	3
2	Low	11	53	35	36	17	16	15
3	Moderate	53	11	28	25	51	23	29
4	High	17	0	1	10	19	29	26
5	Very high	15	0	0	2	0	30	27
Total, %		100	100	100	100	100	100	100

At the same time, a high degree of anthropogenic impact is evident for the majority of Forestries in the European part of the Russian Federation. However, this factor may be critical in conditions of extreme drought, as happened in 2010. So the anthropogenic factor can compensate for low wildfire hazard in terms of fire occurrence in the steppe, forest steppe, and forest zones.

Analysis of the results allows us to conclude that the frequency of fires is determined more by the level of anthropogenic impact on forest lands, and the average area of fires is determined by natural fire danger of forest territories in ways that are consistent with the literature and available data (Mokeev, 1962; Sofronov and Volokitina, 1990).

Classification of forest territory in terms of natural fire hazard and anthropogenic influence is the basis for the organization of the fire monitoring and fire fighting support, and the implementation of preventive measures. It is necessary for successful fire management. In particular, it should be noted that the zoning of Russia in accordance with the type of fire monitoring (see Figure 10.8) is closely correlated with the presented geospatial distributions of natural and anthropogenic fire danger (see Figure 10.9). Hence, on-ground observation is justified for territories with the highest level of anthropogenic impact, even in the promptness with low natural fire hazard and satellite observation is most appropriate for boreal zones with high natural fire risk but low possibility of anthropogenic influence. Despite all this, reasonable preventive measures need to be used to educate the local population in order to improve the culture of behavior in the forests and fire usage.

Local population is the cause of over 80 percent of forest fires in Russia. And the rural population accounts for 72 percent of total wildfires (Andreev & Brukhanov, 2011). According to some statistics these people are mostly aged 20—40 years. About 56—80 percent of the total number of fires is caused by people after vacation or recreation in forests, while 25—36 percent of wildfires are the resulted of human industrial activity in forest (Kuzmichev, 2006; Andreev & Brukhanov, 2011).

In Russia, a number of training events for fire prevention and education of the population have been developed and implemented (Fedorov et al., 2003). These are:

- Forest protective education courses of preschool, high school and other educational institutions (colleges, institutes, universities).
- There are children's "school forestries" in many regions of Russia in the countryside. It is used for children extra education in forestry. Regional and all-Russia conference of "school forestries" members are organized annually.
- Forest fire information and educating through the mass media (posters, flyers, signs forest protection, performance lectures, presentations, and talking on radio and television).
- Installation of volunteer fire brigades.

The current fire situation in Russia, in many respects, reflects the implementation of new forest policy. Reduction was carried out of state forest protection and forest fire specialized units. Forestry has lost about 70,000 forest workers, as well as state Aviation Forest Protection department was decentralized and reduced the number of aircraft and air equipment (Isaev and Korovin, 2013; Goldammer, 2013). This has affected the efficiency of preventing and fighting wildfire including the operational deployment of forces and equipment, monitoring fire detection, and fire fighting air and on ground providing.

ACKNOWLEDGMENTS

This work was supported by a grant of the Russian Science Foundation (project №14-24-00112), and the Russian Foundation for Basic Research.

REFERENCES

Andreev, Y.A., Brukhanov, A.V., 2011. Preventing, Monitoring and Fighting of Forest Fires. Krasnoyarsk (in Russian).

Bartalev, S.A., Yegorov, V.A., Yefremov, V.Y., Loupian, E.A., Stytsenko, D.V., Flitman, E.V., 2012. Integrated burnt area assessment based on combine use of multi-resolution MODIS and Landsat-TM/ETM+ satellite data. Mod. Probl. Remote Sens. 9 (2), 9—26 (in Russian).

Binenko, V.I., Donchenko, V.K., 2004. Damage trends and variability of natural and man-caused environmental disasters. Secur. Emerg. 5, 68—79.

Bondur, V.G., 2011. Satellite monitoring of wildfires in Russia in abnormal drought conditions of 2010. Issled. Zemli Kosmosa (Invest. Earth Space) 3, 3—13 (in Russian).

Conard, S.G., Sukhinin, A.I., Stocks, B.J., Cahoon, D.R., Davidenko, E.P., Ivanova, G.A., 2002. Determining effects of area burned and fire severity on carbon cycling and emissions in Siberia. Clim. Change 55 (1–2), 197–211.

Federal Agency of Forestry, 2008. The main indicators of forest management for 1988, 1992–2007. Wildfire Stat. Available from "Roslesinforg" database.

Fedorov, E.N., Mikhalev, Y.A., Ershova, T.A., 2003. Forest fire propagation as a method of reducing anthropogenic fire hazard. In: Forest Fire Protection, Reforestation and Forest Management, pp. 40–46. Krasnoyarsk.

Vegetation fires and global change: challenges for concerted international action. In: Goldammer, J.G. (Ed.), A White Paper Directed to the United Nations and International Organizations. Global Fire Monitoring Center (GFMC)/Kessel Publishing House.

Isaev, A.S., Korovin, G.N., 2013. Forest as national property of Russia. Lesovedenije 5, 5–12.

Ivanov, V.A., Ivanova, G.A., 2010. Wildfires Caused by Thunderstorms in Siberia. Nauka, Novosibirsk (in Russian).

Kharuk, V.I., Dvinskaya, M.L., Ranson, K.J., 2013. Fire return intervals within the northern boundary of the larch forest in Central Siberia. Int. J. Wildland Fire 22 (2), 207–211. http://dx. doi.org/10.1071/WF11181.

Kudrin, A.Y., Resnikov, V.M., 2006. The system of air and satellite monitoring of wildfire situation. Civ. Secur. Technol. 4 (10), 56–62 (in Russian).

Kukavskaya, E., Soja, A., Petkov, A., Ponomarev, E., Ivanova, G., Conard, Susan, 2013. Fire emissions estimates in Siberia: evaluation of uncertainties in area burned, land cover, and fuel consumption. Can. J. For. Res. 43 (5), 493–506. http://dx.doi.org/10.1139/cjfr-2012-0367.

Kuzmichev, E.P. (Ed.), 2006. Outreach of Population to Prevent Forest Fires. Moscow.

Loupian, E.A., Mazurov, A.A., Flitman, E.V., Ershov, D.V., Korovin, G.N., Novik, V.P., Abushenko, N.A., Altyntsev, D.A., Koshelev, V.V., Tashchilin, S.A., Tatarnikov, A.V., Csiszar, I., Sukhinin, A.I., Ponomarev, E.I., Afonin, S.V., Belov, V.V., Matvienko, G.G., Loboda, T., 2006. Satellite monitoring of forest fires in Russia at federal and regional levels. Mitigation Adapt. Strategies Global Change 11 (1), 113–145.

Mokeev, G.A., 1962. The Principles of Forest Fire Zoning for Regions. Krasnoyarsk (in Russian).

Neigh, C., Nelson, R.F., Ranson, K.J., Margolis, H.A., Montesano, P.M., Sun, G.Q., Kharuk, V.I., Naesset, E., Wulder, M.A., Andersen, H.E., 2013. Taking stock of circumboreal forest carbon with ground measurements, airborne and spaceborne LiDAR. Remote Sens. Environ. 137, 274–287. http://dx.doi.org/10.1016/j.rse.2013.06.019.

Panov, A., Chi, X., Winderlich, J., Birmili, W., Lavrič, J., Ponomarev, E., Andreae, M., 2012. Assessment of biomass burning emissions to the atmosphere from the Zotino Tall Tower Observatory (ZOTTO) in Central Siberia. In: Int. Workshop on Impact of Climate Change on Forest and Agricultural Ecosystems and Adaptation Strategies. Siberian State University, Krasnoyarsk, p. 41.

Ponomarev, E.I., Ivanov, V.A., Korshunov, N.A., 2010. The zoning of forest territory of Russia on the basis of natural fire danger and anthropogenic impact levels. In: Proc. of "VI International Conference on Forest Fire Research". Coimbra, Portugal, 17.

Ponomarev, E.I., Shvetsov, E.G., 2013. Characteristics of vegetation fire categories in siberia, according to satellite-based and other observations. Issled. Zemli Kosmosa (Invest. Earth Space) 5, 45–54 (in Russian).

Ponomarev, E.I., Valendik, E.N., Kisilyakhov, Y.K., 2012. Satellite monitoring of large scale wildfires in siberia. In: Int. Workshop on Impact of Climate Change on Forest and Agricultural Ecosystems and Adaptation Strategies. Siberian State University, Krasnoyarsk, p. 24.

"Rosleskhoz" Official Information, 2013. Retrieved from: http://www.mnr.gov.ru/news/detail.

Rubtsov, A.V., Sukhinin, A.I., Vaganov, E.A., 2011. System analysis of weather fire danger in predicting large fires in Siberian forests. Izv. Atmos. Oceanic Phys. 47 (9), 1049–1056.

Shvidenko, A.Z., Shchepashchenko, D.G., 2013. Climate changes and forest fires in Russia. Lesovedenije 5, 50–61.

Shvidenko, A.Z., Shchepashchenko, D.G., Vaganov, E.A., Sukhinin, A.I., Maksyutov, Sh, McCallum, I., Lakyda, I.P., 2011. Impact of wildfire in Russia between 1998 — 2010 on ecosystems and the global carbon budget. Dokl. Earth Sci. 441 (2), 1678–1682.

Smirnov, A.P., 2013. Forest fires in 2010: causes and effect. Life Saf. J. 11 (Supplement Issue 1. Forest and peat fires), 13–16.

Sofronov, M.A., Volokitina, A.V., 1990. Pyrological Zoning of Taiga. Nauka, Novosibirsk (in Russian).

Soja, A.J., Sukhinin, A.I., Cahoon Jr., D.R., Shugart, H.H., Stackhouse, P.W., 2004. AVHRR-derived fire frequency, distribution and area burned in Siberia. Int. J. Remote Sens. 25 (10), 1939–1960. http://dx.doi.org/10.1080/01431160310001609725.

Sukhinin, A.I., 2008. Space Monitoring and Analysis of Catastrophic Fires in Central Siberia and Far East, vol. 19. Tohoku University, North-East Asia, A la Carte, 19–23.

Sukhinin, A.I., Ivanov, V.V., Ponomarev, E.I., Slinkina, O.A., Cherepanov, A.V., Pavlichenko, E.A., Romasko, V.Y., Miskiv, S.I., 2003. The 2002 fire season in the Asian part of the Russia Federation: a view from space. Int. For. Fire News 28, 18–28.

Vorobjova, Y.L., Akimov, V.A., Sokolov, Y.I., 2004. Forest Fires in Russia: State and Problems. Deka-Press, Moscow.

Chapter 11

Wildland Fire Danger Rating and Early Warning Systems

William J. de Groot
Natural Resources Canada — Canadian Forest Service, Sault Ste. Marie, ON, Canada

B. Michael Wotton
Faculty of Forestry, University of Toronto, Toronto, ON, Canada

Michael D. Flannigan
Dept. of Renewable Resources, University of Alberta, Edmonton, AB, Canada

ABSTRACT

Fire danger rating has become the cornerstone of national fire management programs, and operational systems have been available for over 40 years. Fire danger information is used across a broad spectrum of fire management decision making including daily operations, seasonal strategic planning, and long-term fire and land management planning under future climate change. There are many different national fire danger rating systems in use worldwide. Early warning of extreme fire danger is critical for fire managers to mitigate or prevent wildfire disaster. Early warning is provided using forecasted fire weather, which is further enhanced with remotely sensed fire activity and fuels information in fire early warning systems. Fire danger and early warning systems can operate at global to local levels, depending on fire management requirements. Current operational systems and applications are reviewed.

11.1 INTRODUCTION

Fire danger has been defined as "A general term used to express an assessment of both fixed and variable factors of the fire environment that determine the ease of ignition, rate of spread, difficulty of control, and fire impact; often expressed as an index" (FAO, 2005). Fire danger rating is "A component of a fire management system that integrates the effects of selected fire danger factors into one or more qualitative or numerical indices of current protection needs" (FAO, 2005). Many countries have national fire danger rating systems in place, and

Wildfire Hazards, Risks, and Disasters. http://dx.doi.org/10.1016/B978-0-12-410434-1.00011-7

there are some regional systems that serve numerous countries. Advanced warning of extreme fire danger conditions has become increasingly important in fire management as greater fire activity and disaster fire occurrence has been documented in many global regions (Flannigan et al., 2009). This has occurred for various reasons including climate change, rural—urban population shifts, and land use change affecting vegetation and fuel conditions (Mouillot and Field, 2005; Marlon et al., 2008; Flannigan et al., 2009; Goldammer, 2013). Early warning of impending disaster conditions is a strategy commonly used to minimize or avert the potential devastating impacts of many different natural hazards, and it will become even more important as climate change progresses (Zommers and Singh, 2014). Fire management agencies have also recognized the value of this strategy as fire danger rating systems evolve into fire early warning systems with extended fire weather forecasting and the inclusion of remotely sensed fire, fuels, and weather information.

Wildland fire is similar to other natural hazards in its potential threat to human health and to the safety of many people over large areas, and from the serious postdisaster socioeconomic and environmental impacts it can create (de Groot and Flannigan, 2014; de Groot and Goldammer, 2013). A major difference between wildland fire and other natural hazards is that the occurrence and impact of uncontrolled wildland fire (or wildfire) can be reduced or prevented, at least to a certain degree, through human intervention. Most global wildland fire is caused by humans (Flannigan et al., 2009) so there is great potential to reduce the number of wildland fires on the landscape through fire prevention activities that are initiated by early warning of extreme fire danger conditions. Similarly, early warning can be used to initiate rapid detection and initial attack plans that can prevent human- and lightning-caused fires from quickly spreading to become threatening wildfires. In terms of early warning capacity, fire danger rating systems typically present current (e.g., 4- to 6-h forecast of peak daily burning conditions) or near-term (1- to 3-day forecast) fire danger conditions to support tactical fire management decision making. Some systems provide extended fire danger forecasts (1—2 weeks) and seasonal forecasts (several months) that are produced for longer term strategic planning.

The science of fire danger rating started in Australia, Canada, the United States, and the USSR with research dating back to the 1920s (Chandler et al., 1983). Fire danger rating is used for fire management decision making in all global regions. In northern boreal and temperate forest regions, fire danger is used to measure the daily and hourly change in burning conditions that occur due to highly variable weather. In tropical regions characterized by wet and dry seasons, fire danger rating provides quantitative warning of slowly increasing burning conditions that can reach dangerous levels with little warning. Operational fire danger rating systems have been available for over 40 years, and in that time, fire danger rating has become the cornerstone of day-to-day fire management decision making at the landscape level. Continuing research in this field has also led to more detailed, smaller scale models of fire behavior, fire spread, and fire effects.

11.2 FIRE DANGER RATING SYSTEMS

The need to systematically predict fire activity is common to all jurisdictions where wildfire frequently impacts public safety and natural resources. Through the twentieth century, a number of countries carried out independent research programs that led to the development of danger rating systems for operational use. There are many national fire danger rating systems currently in use worldwide (Table 11.1). The systems range in complexity, reflecting both the complexity of the fire environment and also researchers' observations of the key day-to-day drivers of changing fire activity that need to be understood by fire managers for operational decision making. A brief review of the Nesterov Index from Russia, danger indices developed and used operationally in Australia, and the danger rating systems developed in Canada and the United States provides examples of this range in complexity, and also encompasses the systems most predominantly used in the world. A commonality of each of these systems is that at their core, they rely on estimates of fuel moisture levels on the landscape of interest. In that regard, each system includes (either explicitly or implicitly) an indicator of the rapidly changing moisture content of fine-diameter fuels, which from a very early stage was known to influence fire activity (Show and Kotok, 1925), and also some type of an indicator of drought or the longer term drying of heavier forest fuels. With the exception of the original Nesterov Index, these systems include wind speed as a direct driver of fire danger level, though each system has varying sensitivity to this factor.

11.2.1 The Nesterov Index

The Nesterov Index (Nesterov, 1949; Groisman et al., 2007), used primarily in Russia, is a relatively straightforward function of air temperature, dew point, and rainfall. Its formula can be broken into two basic pieces: the first is a daily increment based on the air temperature—dew point difference multiplied by the air temperature; the second is a cumulative sum of these daily increments since the last day with rain >3 mm. While the cumulative sum is formally used as the Nesterov Index, the daily increment value on its own can be roughly thought of as a relative indicator of the dryness of very fine fuels and thus a good indicator of ease of ignition. The cumulative sum, which resets to 0 when daily rainfall exceeds 3 mm, is then a relative index of cumulative drying of fuels of moderate size class. This complete reset of the index when rainfall exceeds 3 mm can, however, limit it as an indicator of long-term rainfall deficit. A modified version of the Nesterov Index (referred to as the Zhdanko Index) which weighs the accumulation of the daily increment values over a

TABLE 11.1 Characteristics of Common Weather-Based Fire Danger Rating Systems and Indexes[a]

Fire Danger System or Index	Characteristics	Primary Input Parameters	Application Locations	References
Ångström Index	Daily index	Temperature, relative humidity	Sweden, Scandinavia, Germany	Chandler et al. (1983) Holsten et al. (2013)
Baumgartner Index	Cumulative (5-day) index	Precipitation, potential evapotranspiration	Germany, Slovakia	Holsten et al. (2013)
Canadian Forest Fire Weather Index System	Cumulative daily fuel moisture codes; daily fire behavior indices	Temperature, rainfall amount, relative humidity, wind speed	Argentina, Canada, China, Chile, Fiji, Indonesia, Malaysia, Mexico, New Zealand, Portugal, South Africa, Spain, Sweden, Thailand, United Kingdom, United States (Alaska, Florida, north and eastern states), Venezuela; Europe and North Africa, Eurasia, global, Southeast Asia, Southern Africa	Van Wagner (1987) Taylor and Alexander (2006) Wotton (2009)
Fire Danger	Daily index	Temperature, relative humidity, precipitation	Brazil, South America	Setzer and Sismanoglu (2012)

Index	Type	Input variables	Region	References
F Index	Daily indexes of fire danger and fuel moisture	Wind speed, temperature, relative humidity (for Fuel Moisture Index)	United States, Australia, China	Sharples et al. (2009a,b), Liu et al. (2010)
Forest Fire Danger Index	Daily index with cumulative drying factor	Temperature, relative humidity, wind speed, Drought Factor (rainfall, soil dryness, days since rain)	Australia, South Africa, Spain	McArthur (1967), Luke and McArthur (1978)
Fosberg Fire Weather Index	Daily index (with cumulative fuel drying index[b])	Temperature, relative humidity, wind speed, Keetch-Byram Drought Index[b]	United States, Australia, global	Fosberg (1978), Goodrick (2002)
Grassland Fire Danger Index	Daily index	Temperature, relative humidity, wind speed, fuel quantity, degree of grass curing	Australia, South Africa, United States	McArthur (1966), Luke and McArthur (1978)
Haines Index	Daily index	Lower atmosphere temperature and dew point	United States	Haines (1988)
Keetch–Byram Drought Index	Cumulative fuel drying index	Temperature, rainfall amount, mean annual precipitation	United States, Australia, Indonesia	Keetch and Byram (1968), Buchholz and Weidemann (2000)
Lowveld Fire Danger Index	Daily index	Temperature, relative humidity[c]	South Africa	Meikle and Heine (1987)

Continued

TABLE 11.1 Characteristics of Common Weather-Based Fire Danger Rating Systems and Indexes[a]—cont'd

Fire Danger System or Index	Characteristics	Primary Input Parameters	Application Locations	References
National Fire Danger Rating System	Cumulative daily fuel moisture codes; daily fire behavior indices	Temperature, relative humidity, rainfall amount and duration, wind speed, cloudiness, fuel moisture sticks, live fuel moisture (estimated)	United States, South Africa	Deeming et al. (1972, 1977) Bradshaw et al. (1983)
Nesterov Index	Cumulative daily drying index	Temperature, dew point, precipitation amount	Russia, Northern Eurasia, Slovakia, Germany	Nesterov (1949) Buchholz and Weidemann (2000) Groisman et al. (2007) Holsten et al. (2013)
PV 1 or Modified Nesterov Index	Cumulative daily drying index, variable rain factor	Temperature, dew point, precipitation amount	Russia, Canada	Groisman et al. (2007) Ivanova et al. (2011)
Zhdanko Index	Cumulative daily drying index, variable rain factor	Temperature, dew point, precipitation amount	Russia	Zhdanko (1965) Groisman et al. (2007)

[a]There are many other national and subnational systems. For more information see reviews by Viegas et al. (1999), Bovio and Camia (1997), and San-Miguel-Ayanz et al. (2003), and system descriptions by Käse (1969), Sneeuwjagt and Peet (1985), Cheney and Sullivan (1997), Fiorucci et al. (2008), Arpaci et al. (2010), and Won et al. (2010).

[b]Keetch–Byram Drought Index was later incorporated by Goodrick (2002).

[c]A newer Lowveld FDI version for controlled burning includes wind speed, rainfall, and grass curing adjustment factors.

broader range of rainfall values, is also used in some applications (Zhdanko, 1965; Groisman et al., 2007). More recently, a modified Nesterov Index (or PV-1) that includes a precipitation coefficient similar to the Zhdanko Index is also in use (Groisman et al., 2007; Ivanova et al., 2011).

11.2.2 Australian Fire Danger Indices

Fire danger in Australia is estimated most commonly by McArthur's (1967) forest fire danger meter. The version used to create the Forest Fire Danger Index (FFDI), is often referred to as Mark 5, in reference to the version of McArthurs' fire danger meter it is based on. It relies on observed air temperature and relative humidity, wind speed, and a drought factor. McArthur developed the meter as a slide rule tool to be used by fire managers who use the FFDI, but he did not publish his equations along with the slide rule. The slide rule was used to develop a set of equations for the FFDI calculation (Noble et al., 1980). Like the Nesterov Index, the McArthur fire danger meter calculation does not explicitly estimate the moisture content of the fuels sustaining flaming combustion, but implicitly accounts for these through the inclusion of temperature and relative humidity in the danger index estimation. Viney (1992), however, developed an equation for fuel moisture from information in McArthur's (1967) publication. The Drought Factor component of the FFDI calculation, which is based on the Keetch–Byram Drought Index (Keetch and Byram, 1968), can be thought of as an indicator of the availability of fuels for combustion based on the assumption that increasing drought will lead to increased dead fuels being made available for consumption. The operational scale of the FFDI ranges from 0 to 100, although recent extreme fire weather and consequent catastrophic fire activity has led to FFDI values higher than 100 (Cruz et al., 2012). McArthur (1966) also produced a grassland fire behavior meter which is used throughout much of Australia. Currently, the grassland meter uses air temperature, relative humidity, and wind speed along with the degree of grassland curing and fuel load to estimate a Grassland Fire Danger Index similar to the FFDI.

In Western Australia, fire danger is also estimated for use in preparedness planning by the Forest Fire Behavior Tables (FFBT) (Sneeuwjagt and Peet, 1985; Beck, 1995) which produce an index for several fire-prone fuel complexes in the region. This index is a function of wind speed and litter moisture content, the latter being estimated by a bookkeeping method. Matthews (2009) provides a short review of McArthur's FFDI and the FFBT (and also the Canadian Fire Weather Index (FWI)) providing equations and comparisons of each index for a site in Australia.

11.2.3 The Canadian Forest Fire Weather Index System

Dedicated fire research in Canada began in the late 1920s with the design of a comprehensive field-based fuel moisture monitoring and ignition sustainability testing research program. Fire hazard and fire danger tables designed for use by fire suppression agencies quickly followed and were adopted into operational use in different regions of the country. As a response to the growing number of regional fire hazard and danger tables, a universal national system was conceptualized in 1967 (Muraro, 1968). The system evolved into the currently used Canadian Forest Fire Weather Index (FWI) System in the late 1960s and its structure has remained relatively unchanged. The FWI System is a set of six elements each describing either the state of fuel moisture in differing fuel layers or some aspect of fire behavior potential. Like the previously discussed fire danger rating systems, the FWI System is a weather-based method that tracks fire danger in a single standard fuel type. The standard fuel type is a closed canopy mature jack pine stand, which is commonly found throughout the boreal forest, although it is by no means the only important fuel type in Canada. However, outputs of the FWI System are used as relative indicators of fire potential across all forest types in the country, with local experience of fire management personnel used to calibrate index values to observed outcomes.

Inputs of the FWI System are air temperature, relative humidity, wind speed, and rainfall. The first three outputs of the system are moisture codes, which are bookkeeping systems tracking moisture in three important layers of the forest floor: (1) surface litter using the Fine Fuel Moisture Code (FFMC), (2) the upper soil organic layer (also used to represent medium-sized dead and downed woody material, or branchwood) using the Duff Moisture Code (DMC), and (3) the deeper soil organic layer (also used to represent heavy dead and downed woody material, or logs) using the Drought Code (DC). Within the FWI System, moisture content values are transformed to "moisture codes" so that high code values represent low moisture content and thus increasing code values indicate an increasing fire danger situation.

The final three main elements of the FWI System are unitless fire behavior indexes, each representing an element of Byram's classic fire line intensity formulation (Byram, 1959). The Initial Spread Index (ISI) is a relative indicator of the potential spread rate of a fire, through a combination of wind speed and litter moisture (as estimated by the FFMC). The Buildup Index (BUI) is a relative indicator of fuels available for combustion as it represents a combination of the two heavier fuel moisture codes from the system, the DMC and DC. Finally, the FWI is a relative indicator of head fire intensity. The FWI is a multiplicative combination of ISI and BUI (similar to Byram's fire line intensity $I = HWR$, where the ISI is a surrogate for R (rate of fire spread) and the BUI is a surrogate for W (fuel consumed); H is heat of combustion) that has been transformed so that it scales in a similar fashion to flame length. Early fire danger indexes in Canada (from the middle of the twentieth century) reported danger on a scale from 1 to 16; however, in modern implementation of the system, the FWI is, mathematically, unlimited.

It is important to make the clear distinction that while the FWI is one of six main indexes of the system, the term "FWI System" is used to describe the entire system of codes and indexes. While it is the FWI that is usually used to inform public warnings and set the level of roadside fire danger signs, modern fire management agencies use each element of the FWI System in helping formulate their management decisions. For instance, the FFMC is the system's best indicator of the potential for human-caused fires, the DMC is the system's best indicator for the receptivity of the forest to ignition from lightning, and the BUI is a good indicator of the potential difficulty of suppression of a fire, as it represents the effect of long-term rainfall deficit on fuels (with increasing BUI values reflecting a high availability of consumable fuel). The main documentation describing the development and structure of the FWI System is Van Wagner (1987), while Van Wagner and Pickett (1985) provide a useful summary of the equations in a concise form. Recent reviews describing the FWI System in relation to fire management and research have been published by Taylor and Alexander (2006) and Wotton (2009).

11.2.4 The US National Fire Danger Rating System

Fire research in the United States began soon after the formation of the US Forest Service with Gisborne (1929) defining the key elements contributing to fire danger: fuel moisture, wind, and relative humidity. Similar to the Australian experience, this early research led to development of a fire danger meter, which was subsequently improved upon in various revisions over the following two and a half decades. By the late 1950s, numerous regionally specific fire danger meters had developed across the United States (similar to the experience in Canada), which led to the proposal of a single national fire danger rating system, which ultimately led to the development of Rothermel's (1972) fire spread model that was adopted as the fundamental fire behavior driver underlying the National Fire Danger Rating System (NFDRS). First, published in 1972 (Deeming et al., 1972), the NFDRS was updated in 1978 (Deeming et al., 1977; Bradshaw et al., 1983) and in 1988, the latter time mainly to improve performance in parts of the eastern and southeastern United States.

Like the Canadian FWI System, the NFDRS has a number of sub-components and indexes used in the danger rating estimation process but it uses a greater number of inputs and has a greater number of subcomponents or indexes. Like the other systems described earlier, the NFDRS uses basic weather observations (air temperature, relative humidity, wind speed, and rainfall) as inputs to drive core parts of the overall system; however, these weather variables form only one part of the input. Additional weather requirements for the calculations are cloudiness, minimum and maximum temperature and relative humidity, and precipitation duration (in addition to the standard precipitation amount). These weather inputs are used to estimate moisture content of different-sized fuels important for combustion.

An actual estimate of moisture derived from fuel moisture sticks, arrays of pine dowels weighed each day to assess their moisture content, is also an input.

While this manual observation of daily stick moisture was the original source of these observations, in recent years electronic sticks and weather-based stick moisture models have been developed to provide additional methods for estimating this required value. The system also requires an estimate of live fuel moisture. With the exception of this live fuel moisture estimate, the US danger rating system does not track moisture in specific layers of the fuel complex (i.e. the litter layer) but relies upon estimated moisture in a range of fuel sizes which correspond to fuel element time lags (1 h, 10 h, 100 h, 1,000 h) and represent increasingly heavier (larger diameter) fuels. This allows users of the system to estimate the effective moisture content of a range of fuel complexes. Correspondingly the system defines a set of fuel models for which each of its final fire behavior indices can be calculated.

The main outputs of the system, from a danger rating perspective, are similar to the final three outputs of the FWI System. The Spread Component (SC) is the spread rate from the Rothermel model. Unlike the Canadian system's ISI, this element of the NFDRS has physical units: ft/min. The Energy Release Component (ERC) represents the energy flux from the main head of the fire, and is influenced by the amount of fuel present and also the size distribution of the elements of that fuel complex. The SC and ERC combine to produce the final Burning Index (BI) which is scaled such that it represents about 10 times the expected flame length of a fire. It is therefore a good general indicator of suppression difficulty. While the BI can be thought of as the main summary "Fire Danger Index" from the system (analogous to the FWI in the Canadian system, the FFDI in the Australian system, and the Nesterov Index), operational fire managers use each of the output components in different aspects of the their decision-making process.

11.3 FIRE EARLY WARNING SYSTEMS

11.3.1 Temporal Scale in Warning Systems

Fire danger rating forms the basis of early warning systems in wildland fire management and there are different levels of early warning that serve different purposes (de Groot and Flannigan, 2014). For example, a community that needs to be evacuated due to imminent wildfire threat requires people to be informed as quickly as possible. In the worst cases, early warning requirements could be on the order of minutes to a few hours. This is more accurately described as "very short term" early warning. In recent years, there has been increasing use of social media by fire management agencies to quickly provide fire information to the public. The Advanced Fire Information System (AFIS) in South Africa is an example of a system that quickly informs fire and land managers of changing local fire situation (Davies et al., 2008). AFIS has been operational since 2004 and it provides near-real-time warning of active fires and other fire-related information to desktop and cell phones. AFIS warnings are based on Moderate

Resolution Imaging Spectroradiometer (MODIS) and Meteosat Second Generation (MSG) satellite-detected hot spots and the user-selected location.

At a slightly longer time frame, "short-term" early warning of hourly fire danger is often provided to subnational levels, or for small-scale areas such as ongoing fires, during times of extreme fire danger or when fire danger is rapidly changing. This information is very important for fire management decisions related to suppression planning on campaign fires and preparing for new fire starts during periods of extreme burning conditions. "Medium-term" early warning of peak daily fire danger that is provided several days in advance is typical at national and larger spatial scales. This is the most commonly used early warning information, and it supports most day-to-day fire management decision making. National, regional, and global scales also use "long-term" early warning of daily fire danger that is days to weeks in advance to make large-scale decisions such as planning national or international mobilization and distribution of suppression resources. Seasonal forecasts of expected future fire danger trends that are provided months in advance can be used as early warning for long-term planning, such as anticipated seasonal budgeting.

Fire danger rating and early warning systems can operate at any spatial scale depending on the size of the jurisdiction and system purpose. In general, there are four possible operating levels: global, regional, national, and subnational (including local). At larger scales, information is usually based on coarse resolution data (spatial and temporal scale), while systems operating at smaller scales use more detailed and frequent local data. Therefore, fire managers need to identify the level of information required for the purpose intended, accepting that there is some necessary compensation in scale.

11.3.2 Spatial Scale in Warning Systems

11.3.2.1 The Global Early Warning System for Wildland Fire

Following the United Nations World Conference on Disaster Reduction (Kobe, Japan, January 2005), the Global Early Warning System for Wildland Fire (EWS-Fire) was initiated as a project under the Fire Implementation Team of the Global Observation of Forest and Land Cover Dynamics (GOFC-GOLD) in support of the Hyogo Framework for Action priorities (UNISDR, 2007). The impetus for a unified global system was recognition by the international wildland fire community that important keys to reducing global wildfire disaster was early warning and increased international cooperation between wildland fire agencies (FAO, 2006). The purpose of the Global EWS-Fire is to support the many different national fire danger rating systems in use and enhance international fire management cooperation by providing a common international metric for implementing international resource sharing agreements during times of fire disaster, new longer term predictions of fire danger based on advanced numerical weather models, and a fire danger rating system for the many countries that do not have the financial or institutional capacity to develop a national system (de Groot and Goldammer, 2013).

The Global EWS-Fire is being developed to be able to provide a variety of early warning products that support decision making across the fire management spatial and temporal scales (de Groot and Goldammer, 2013) from long-term, large-scale fire control strategy such as international fire suppression resource sharing, to local day-to-day community-based fire management activities.

Having a single fire danger rating system for international use will provide a common method to assess burning conditions across countries and global regions that operate many different national systems. This will facilitate a more holistic understanding of the larger scale fire environment and the fire management situation of other nations. For individual countries, it will be much simpler to conduct and interpret one cross-calibration between the national system and one international system, than to cross-calibrate with many national systems. Extending global fire danger forecasts with data from longer term forecast weather models will provide greater early warning capacity which would facilitate coordinatation of suppression resource sharing and mobilization within and between countries in advance of disaster conditions. The global system can also be used as a basic national system in the many countries that do not have a system in place, which currently represents over half of the world's nations. Although the Global EWS-Fire is at a coarse scale, it can support local fire danger information and community-based fire management programs (FAO, 2011) by indicating future fire danger trends.

Still in a relatively early stage of development, the Global EWS-Fire is envisioned to be a tool that will continually evolve as new science and data sources become available. The system currently displays an interactive series of 1- to 7-day forecasted global-level fire danger maps using the FWI System; the user is able to overlay current hot spots and national boundaries and also more closely examine specific areas of interest (Figure 11.1). Currently, accessed through the Global Fire Monitoring Center [1], it is anticipated that in the future the Global EWS-Fire system will be accessible through several different portals related to the contributing agencies (e.g., Joint Research Centre of the European Commission, Desert Research Institute, and Natural Resources Canada—Canadian Forest Service) as well as the GOFC-GOLD Fire Implementation Team Web site.

11.3.2.2 Regional Systems

Regional fire danger rating/early warning systems are an efficient way to operate a common system for many countries in an area with similar land and forest values and fire issues. International resource sharing is most frequent between countries within the same global region, so regional systems can be a very important tool for enhancing international fire management collaboration.

---

1. Global EWS-Fire website: www.fire.uni-freiburg.de/gwfews/index.html.

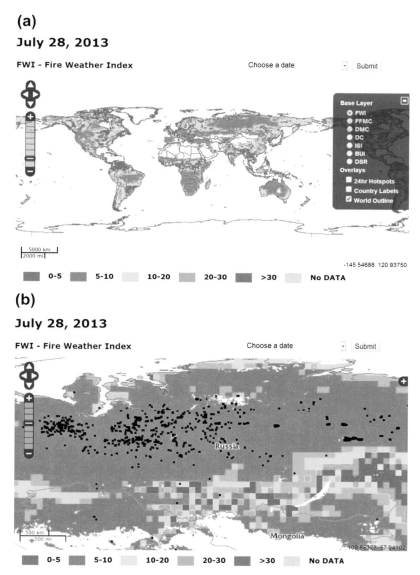

FIGURE 11.1   Examples of global early warning system for wildland fire (EWS-fire) products at the (a) global and (b) regional level (with current hot spots displayed in black).

Regional systems also generally provide more detailed fire danger and early warning information than a global system because they use higher density weather data networks.

There are several regional fire danger rating/early warning systems that are currently in operation. The Malaysia Meteorological Department operates a regional system for the 10 countries of the Association of Southeast Asian Nations. The South East Asia Fire Danger Rating System is a regionally calibrated version of the FWI System and was developed following the 1997−1998 haze disaster in Southeast Asia (similar national systems were developed for Indonesia and Malaysia, which operate independently) (de Groot et al., 2007a). The Joint Research Center of the European Commission operates the European Forest Fire Information System (EFFIS) (San-Miguel-Ayanz et al., 2002), which provides daily fire danger to all European countries and northern Africa using the FWI System. In this case, EFFIS provides a common fire danger rating system to a region where there are many countries using different national systems and also some countries without a national system. Brazil's National Institute for Space Research produces daily fire danger maps for South America using a locally derived Fire Danger Index. Finally, South Africa's Council for Scientific and Industrial Research operates the Wide Area Monitoring Information System, which presents daily and forecasted fire danger for southern Africa using the Lowveld Fire Danger Index and the Fire Weather Index component of the FWI System.

11.3.2.3 National Systems

The primary decision-making point in centrally organized fire management agencies is at the national or subnational (province, state, or territory) levels. In decentralized organizations, primary decision making generally occurs at subnational or local levels. National fire danger rating system products are often provided for fire managers at national or smaller scales (Figure 11.2). There are many national fire danger rating systems currently in use around the world (Table 11.1) with varying degrees of early warning capacity. At national or smaller scale levels, fire danger information is more detailed (greater density of national synoptic or fire weather station networks) and often more frequently updated (sometimes multiple times per day, or even hourly during extreme conditions), but is also representative of much smaller areas than global or regional systems. Early warning capacity usually decreases at small-scale levels because of limited access to forecast weather models and remotely sensed data, although flexibility in obtaining the most current fire danger information is greatest at the lowest levels. For instance, at the local level (e.g., community, district, park), fire danger calculations can be updated an unlimited number of times each day using local weather station data, which is very valuable for making immediate tactical decisions (e.g., mobilizing suppression resources to various locations at ongoing fires) even during times of significant weather change. However, early warning information for making longer term strategic decisions (e.g., increasing or decreasing overall suppression resource levels) in anticipation of changing fire danger conditions is usually available at national or higher levels.

(a)

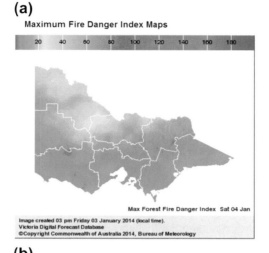

(b)

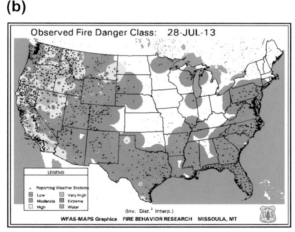

FIGURE 11.2   Examples of spatial fire danger products for (a) Australia's Forest Fire Danger Index (FFDI) for the State of Victoria and (b) the US National Fire Danger Rating System.

11.4 FORECASTING FIRE DANGER FOR EARLY WARNING

For all the weather-based fire danger rating systems in Table 11.1, early warning depends on forecasting atmospheric conditions that affect fuel moisture and fire behavior. The atmosphere behaves according to a set of physical laws often referred to as the equations of motion. These equations can represent numerous atmospheric variables including temperature, pressure, and wind speed at the surface and at various levels above the Earth. These equations describe also how variables will change over time so that we can forecast or predict the future state of the atmosphere. There are no exact analytical solutions to these equations so numerical methods are employed to make these forecasts of future weather. There are numerous Meteorological Organizations around the world that run these Numerical Weather Prediction (NWP) computer programs to routinely forecast the weather at various spatial scales from local, regional, national, to global, and for temporal scales from subhourly to 15 days and even longer. These forecasts are typically produced once or twice daily to coincide with globally coordinated observations of meteorological data from surface and upper air stations (upper air stations release radiosondes that collect data at mandatory and significant levels in the atmosphere) at 00 and 1,200 GMT. This means a new weather forecast is produced every 12 h that facilitates a new fire danger forecast every 12 h. As a result, an early warning system for fire danger rating can be continually updated.

In an early warning system, it is the weather that is most dynamic as the fuels are generally considered static, although vegetation can green up over a relatively short period. Fuel moisture, a critical aspect for fire danger rating and early warning systems, is largely determined by the weather and this will also influence ignition probabilities. Many fire danger rating systems use meteorological variables such as temperature, precipitation, relative humidity, and wind speed (Table 11.4) along with fuel type to estimate fuel moisture, rate of spread, and fire intensity. With global gridded data of these meteorological variables, it is possible to predict the fire weather and fire danger for 15 days or longer into the future. The accuracy of these NWP forecasts is typically measured against climatology and persistence (tomorrow's forecast being the same as today's observations). The accuracy of most models lasts for 5–10 days depending on the model and the atmospheric situation. The best forecasts are often achieved using an ensemble approach that uses an average of many forecasts from different models, different parameterizations, or by making small changes (perturbations) in the initial conditions (Krishnamurti et al., 2000). Additionally, seasonal weather forecasts are available using NWP outputs or using other approaches based on sea surface temperature–climate oscillations (e.g., El Niño Southern Oscillation) and these seasonal weather forecasts can generate a fire season forecast (Roads et al., 2005; Shabbar et al., 2011). This seasonal forecast is more of a climate rather that a weather forecast, as it will give an indication of the severity of the fire season but does not provide useful day-to-day fire weather/fire danger forecasts for the upcoming fire season.

11.5 FIRE DANGER AND EARLY WARNING APPLICATIONS

Fire danger rating and early warning form the foundation for decision making in contemporary fire management programs (Taylor and Alexander, 2006; Wotton, 2009; de Groot and Goldammer, 2013). Fire danger information is typically applied in fire management through various decision aid tools and standard operating procedures (SOPs) that are specifically tailored to an agency's fire management program and the local fire environment. Decision aids are usually developed using agency-specific fire environment parameters (historical fire and weather statistics, fuels data) that are linked to local fire management criteria (policy, funding, suppression resources, infrastructure, etc.). Various methods such as fire behavior modeling (Andrews, 1986; Forestry Canada Fire Danger Group, 1992) and/or statistical modeling are used to provide the linkage between physical fire environment and fire management. Decision aids are often quantitative to simplify decision making. SOPs use decision aid criteria to trigger agency-based action plans for fire prevention, detection, presuppression resource mobilization, etc. (cf. Figure 7.3 in de Groot and Flannigan, 2014). Fire-danger-based decision aids are also used for long-term fire management planning including prescribed burn planning (Andrews, 2014) and hazard reduction (Fernandes and Botelho, 2003), predicting fire effects (Keane et al., 1989; Albini and Reinhardt, 1995; Albini et al., 1995; Ottmar et al., 1993; Reinhardt et al., 1997; Reinhardt and Crookston, 2003; de Groot, 2010), modeling landscape fire and vegetation dynamics using physical and/or weather-based landscape models (c.f. Keane et al., 2004, 2013), and fire and carbon emissions modeling (de Groot et al., 2007b; French et al., 2011). Fire danger models also serve as the basis for estimating the impacts of climate change on future fire regimes (Wotton et al., 2010; de Groot et al., 2013; Flannigan et al., 2013) and future fire management (Podur and Wotton, 2010), as well as informing other research studies that support fire management decision making (Wotton, 2009). Research examples include correlating fire danger with area burned (Abatzoglou and Kolden, 2013) and large fire occurrence (Riley et al., 2013), modeling fire suppression costs (Holmes and Calkin, 2013; Houtman et al., 2013), building ignition prediction models (Wotton and Martell, 2005), managing landscape fuels (Finney et al., 2007), assessing landscape wildfire risk (Finney et al., 2011), managing wildfire risk at the wildland—urban interface (Bradstock et al., 1998; Price and Bradstock, 2012), and modeling fire suppression effectiveness (Podur and Martell, 2007; Plucinski and Pastor, 2013).

11.6 FUTURE FIRE DANGER AND EARLY WARNING

There are a number of enhancements in fire danger rating and early warning information that can be expected in the future. Spatial mapping of precipitation (using radar and satellite data) would significantly improve spatial accuracy of fire danger maps, particularly in areas and seasons where smaller scale convective activity can dominate rainfall patterns. As weather-driven fire behavior models improve, there will be a requirement for more detailed fuel mapping, including fuel structure and fuel load information, which could be provided with frequent updates through remote sensing. Also, remote monitoring of live fuel moisture (Ceccato et al., 2003) will assist in determining landscape-level fuel flammability and seasonal criteria that are important to predicting fire behavior.

ACKNOWLEDGMENTS

Information on individual fire danger rating systems provided by Johann G. Goldammer, Harald Vacik, Klaus-Peter Wittich, Paolo Fiorucci, and Francesco Gaetani is gratefully appreciated. Alan Cantin prepared the manuscript figures.

REFERENCES

Abatzoglou, J.T., Kolden, C.A., 2013. Relationships between climate and macroscale area burned in the western United States. Int. J. Wildland Fire 22, 1003–1020.

Albini, F.A., Brown, J.K., Reinhardt, E.D., Ottmar, R.D., 1995. Calibration of a large fuel burnout model. Int. J. Wildland Fire 5, 173–192.

Albini, F.A., Reinhardt, E.D., 1995. Modeling ignition and burning rate of large woody natural fuels. Int. J. Wildland Fire 5, 81–91.

Andrews, P.L., 1986. BEHAVE: Fire Behavior Prediction and Fuel Modelling System–BURN Subsystem, Part I. USDA Forest Service, Ogden, UT. Gen. Tech. Rep. INT-194.

Andrews, P.L., 2014. Current status and future needs of the BehavePlus fire modeling system. Int. J. Wildland Fire 23, 21–33.

Arpaci, A., Vacik, H., Formayer, H., Beck, A., 2010. Alpine Forest Fire Warning System, a Collection of Possible Fire Weather Indices (FWI) for Alpine Landscapes, 29 pp. (unpub. report, Alpine Forest Fire Warning System Project). http://www.alpffirs.eu/.

Beck, J.A., 1995. Equations for the forest fire behaviour tables for western Australia. CALM Sci. 1, 325–348.

Bovio, G., Camia, A., 1997. Meteorological indices for large fires danger rating. In: Chuvieco, E. (Ed.), A Review of Remote Sensing Methods for the Study of Large Wildland Fires. Universidad de Alcalà, Alcalà de Henares, Spain, pp. 73–91.

Bradshaw, L.S., Deeming, J.E., Burgan, R.E., Cohen, J.D., 1983. The National Fire Danger Rating System: Technical Documentation. USDA For. Serv., Gen. Tech. Rep. INT-169, Ogden, UT.

Bradstock, R.A., Gill, A.M., Kenny, B.J., Scott, J., 1998. Bushfire risk at the urban interface estimated from historical weather records: consequences for the use of prescribed fire in the Sydney region of south–eastern Aust. J. Environ. Manag. 52, 259–271.

Buchholz, G., Weidemann, D., 2000. The use of simple fire danger rating systems as a tool for early warning in forestry. Int. For. Fire News 23, 32–36.

Byram, G.M., 1959. Combustion of forest fuels. In: Davis, K.P. (Ed.), Forest Fire: Control and Use. McGraw-Hill, New York, pp. 61–89.

Ceccato, P., Leblon, B., Chuvieco, E., Flasse, S., Carlson, J., 2003. Estimation of live fuel moisture content. In: Chuvieco, E. (Ed.), Wildland Fire Danger Estimation and Mapping - the Role of Remote Sensing. World Scientific Publishing, Singapore, pp. 63–90.

Chandler, C., Cheney, P., Thomas, P., Trabaud, L., Williams, D., 1983. Fire in Forestry. In: Forest fire Behavior and Effects, vol. 1. John Wiley & Sons, New York.

Cheney, N.P., Sullivan, A., 1997. Grassfires: Fuel, Weather and Fire Behaviour. CSIRO Publishing, Canberra.

Cruz, M.G., Sullivan, A.L., Gould, J.S., Sims, N.C., Bannister, A.J., Hollis, J.J., Hurley, R.J., 2012. Anatomy of a catastrophic wildfire: the black saturday Kilmore east fire in Victoria, Australia. For. Ecol. Manag. 284, 269–285.

Davies, D.K., Vosloo, H.F., Frost, P.E., Vannan, S.S., 2008. Near real-time fire alert system in South Africa: from desktop to mobile service. In: Proceedings of the 7th ACM Conference on Designing Interactive Systems. ACM, Cape Town, pp. 315–322.

Deeming, J.E., Lancaster, J.W., Fosberg, M.A., Furman, W.R., Schroeder, M.J., 1972. The National Fire-Danger Rating System. Res. Pap. RM-84. USDA Forest Service, Fort Collins, CO (revised 1974).

Deeming, J.E., Burgan, R.E., Cohen, J.D., 1977. The National Fire Danger Rating System – 1978. USDA Forest Service, Ogden, UT. Gen. Tech. Rep. INT-39.

de Groot, W.J., 2010. Modeling fire effects: integrating fire behavior and fire ecology. In: Viegas, D.X. (Ed.), 6th International Conference on Forest Fire Research. ADAI/CEIF University of Coimbra, Coimbra, Portugal.

de Groot, W.J., Goldammer, J.G., 2013. The global early warning system for wildland fire. In: Goldammer, J.G. (Ed.), Vegetation Fires and Global Change - Challenges for Concerted International Action. A White Paper Directed to the United Nations and International Organizations. Kessel Publishing House, Remagen-Oberwinter, Germany, pp. 277–284.

de Groot, W.J., Field, R.D., Brady, M.A., Roswintiarti, O., Mohamad, M., 2007a. Development of the Indonesian and Malaysian fire danger rating systems. Mitig. Adapt. Strat. Glob. Change 12, 165–180.

de Groot, W.J., Flannigan, M.D., 2014. Climate change and early warning systems for wildland fire. In: Zommers, Z., Singh, A. (Eds.), Reducing Disaster: Early Warning Systems for Climate Change. Springer, New York, pp. 127–151.

de Groot, W.J., Flannigan, M.D., Cantin, A.S., 2013. Climate change impacts on future boreal fire regimes. For. Ecol. Manag. 294, 35–44.

de Groot, W.J., Landry, R., Kurz, W.A., Anderson, K.R., Englefield, P., Fraser, R.H., Hall, R.J., Banfield, E., Raymond, D.A., Decker, V., Lynham, T.J., Pritchard, J.M., 2007b. Estimating direct carbon emissions from Canadian wildland fires. Int. J. Wildland Fire 16, 593–606.

Gisborne, H.T., 1929. The complicated controls of fire behaviour. J. For. 27, 311–312.

FAO, 2005. Wildland Fire Management Terminology. Food and Agriculture Organization, United Nations, Rome (Updated Jan 2005). TRG(A10.6)/GICM. http://fao.org/forestry/firemanagement/13530/en/ (accessed 7.01.14.).

FAO, 2006. Fire Management Review: International Cooperation. Fire Manage. Work. Pap. 18. Food and Agriculture Organization, United Nations, Rome.

FAO, 2011. Community-Based Fire Management. For. Pap. 166. Food and Agriculture Organization, United Nations, Rome.

Fernandes, P.M., Botelho, H.S., 2003. A review of prescribed burning effectiveness in fire hazard reduction. Int. J. Wildland Fire 12, 117–128.

Finney, M.A., Seli, R.C., McHugh, C.W., Ager, A.A., Bahro, B., Agee, J.K., 2007. Simulation of long-term landscape-level fuel treatment effects on large wildfires. Int. J. Wildland Fire 16, 712–727.

Finney, M.A., McHugh, C.W., Grenfell, I.C., Riley, K.L., Short, K.C., 2011. A simulation of probabilistic wildfire risk components for the continental United States. Stoch. Environ. Res. Risk Assess. 25, 973–1000.

Fiorucci, P., Gaetani, F., Minciardi, R., 2008. Development and application of a system for dynamic wildfire risk assessment in Italy. Environ. Model. Softw. 23, 690–702.

Flannigan, M.D., Cantin, A.S., de Groot, W.J., Wotton, M., Newbery, A., Gowman, L.M., 2013. Global wildland fire severity in the 21st century. For. Ecol. Manag. 294, 54–61.

Flannigan, M.D., Krawchuk, M.A., de Groot, W.J., Wotton, B.M., Gowman, L.M., 2009. Implications of changing climate for global wildland fire. Int. J. Wildland Fire 18, 483–507.

Forestry Canada Fire Danger Group, 1992. Development and Structure of the Canadian Forest Fire Behavior Prediction System. Inf. Rep. ST-X-3, Forestry Canada, Ottawa.

Fosberg, M.A., 1978. Weather in wildland fire management: the fire weather index. In: Proceedings of the Conference on Sierra Nevada Meteorology. American Meteorological Society, Boston, pp. 1–4.

French, N.H.F., de Groot, W.J., Jenkins, L.K., Rogers, B.M., Alvarado, E., Amiro, B., de Jong, B., Goetz, S., Hoy, E., Hyer, E., Keane, R., Law, B.E., McKenzie, D., McNulty, S.G., Ottmar, R., Pérez-Salicrup, D.R., Randerson, J., Robertson, K.M., Turetsky, M., 2011. Model comparisons for estimating carbon emissions from north American wildland fire. J. Geophys. Res. 116, G00K05.

Goldammer, J.G. (Ed.), 2013. Vegetation Fires and Global Change - Challenges for Concerted International Action. A White Paper Directed to the United Nations and International Organizations. Kessel Publishing House, Remagen-Oberwinter, Germany.

Goodrick, S.L., 2002. Modification of the Fosberg fire weather index to include drought. Int. J. Wildland Fire 11, 205–211.

Groisman, P.Y., Sherstyukov, B.G., Razuvaev, V.N., Knight, R.W., Enloe, J.G., Stroumentova, N.S., Whitfield, P.H., Førland, E., Hannsen-Bauer, I., Tuomenvirta, H., Aleksandersson, H., Mescherskaya, A.V., Karl, T.R., 2007. Potential forest fire danger over northern Eurasia: changes during the 20th century. Glob. Planet. Change 56, 371–386.

Haines, D.A., 1988. A lower atmospheric severity index for wildland fire. Nat. Weather Dig. 13, 23–27.

Holmes, T.P., Calkin, D.E., 2013. Econometric analysis of fire suppression production functions for large wildland fires. Int. J. Wildland Fire 22, 246–255.

Holsten, A., Dominic, A.R., Costa, L., Kropp, J.P., 2013. Evaluation of the performance of meteorological forest fire indices for German federal states. For. Ecol. Manag. 287, 123–131.

Houtman, R.M., Montgomery, C.A., Gagnon, A.R., Calkin, D.E., Dieterich, T.G., McGregor, S., Crowley, M., 2013. Allowing a wildfire to burn: estimating the effect on future fire suppresson costs. Int. J. Wildland Fire 22, 871–882.

Ivanova, G.A., Conard, S.G., Kukavskaya, E.A., McRae, D.J., 2011. Fire impact on carbon storage in light conifer forests of the lower Angara region, Siberia. Environ. Res. Lett. 6, 045203. http://dx.doi.org/10.1088/1748–9326/6/4/045203.

Käse, H., 1969. Ein Vorschlag für eine Methode zur Bestimmung und Vorhersage der Waldbrandgefährdung mit Hilfe komplexer Kennziffern, vol. 68. Akademie Verlag, Berlin.

Keane, R.E., Arno, S.F., Brown, J.K., 1989. FIRESUM – an Ecological Process Model for Fire Succession in Western Conifer Forests. USDA Forest Service, Ogden, UT. Gen. Tech. Rep. INT-266.

Keane, R.E., Cary, G.J., Davies, I.D., Flannigan, M.D., Gardner, R.H., Lavorel, S., Lenihan, J.M., Li, C., Rupp, T.S., 2004. A classification of landscape fire succession models: spatial simulations of fire and vegetation dynamics. Ecol. Model. 179, 3–27.

Keane, R.E., Cary, G.J., Flannigan, M.D., Parsons, R.A., Davies, I.D., King, K.J., Li, C., Bradstock, R.A., Gill, M., 2013. Exploring the role of fire, succession, climate, and weather on landscape dynamics using comparative modeling. Ecol. Model. 266, 172–186.

Keetch, J.J., Byram, G., 1968. A Drought Index for Forest Fire Control. Res. Pap. SE-38. USDA Forest Service, Asheville, NC (revised 1988).

Krishnamurti, T.N., Kishtawal, C.M., Zhang, Z., LaRow, T., Bachiochi, D., Williford, E., Gadgil, S., Surendran, S., 2000. Multimodel ensemble forecasts for weather and seasonal climate. J. Clim. 13, 4196–4216.

Liu, X., Zhang, J., Cai, W., Tong, Z., 2010. Information diffusion-based spatio-temporal risk analysis of grassland fire disaster in northern China. Knowledge-Based Syst. 23, 53–60.

Luke, R.H., McArthur, A.G., 1978. Bushfires in Australia. Australian Government Public Service, Canberra.

Marlon, J.R., Bartlein, P.J., Carcaillet, C., Gavin, D.G., Harrison, S.P., Higuera, P.E., Joos, F., Power, M.J., Prentice, I.C., 2008. Climate and human influences on global biomass burning over the past two millennia. Nat. Geosci. 1, 697−702.

Matthews, S., 2009. A comparison of fire danger rating systems for use in forests. Aust. Meteorol. Oceanogr. J. 58, 41−48.

McArthur, A.G., 1966. Weather and Grassland Fire Behaviour. Department of National Development, Forestry and Timber Bureau, Canberra. Leaflet No. 100.

McArthur, A.G., 1967. Fire Behaviour in Eucalypt Forests. Department of National Development, Forestry and Timber Bureau, Canberra. Leaflet No. 107.

Meikle, S., Heine, J., 1987. A fire danger index system for the Transvaal Lowveld and adjoining escarpment area. S. Afr. For. J. 143, 55−56.

Mouillot, F., Field, C.B., 2005. Fire history and the global carbon budget: a $1° \times 1°$ fire history reconstruction for the 20th century. Glob. Change Biol. 11, 398−420.

Muraro, S.J., 1968. A modular approach to a revised national fire danger rating system. In: Contributions to the Development of a National Fire Danger Rating System. Can. For. Serv., Inf. Rep. BC-X-37, Victoria, BC.

Nesterov, V.G., 1949. Combustibility of the Forest and Methods for Its Determination. USSR State Industry Press, Moscow, Goslesbumizdat, 76 pp. (In Russian).

Noble, I.R., Bary, G.A.V., Gill, A.M., 1980. McArthur's fire-danger meters expressed as equations. Aust. J. Ecol. 5, 201−203.

Ottmar, R.D., Burns, M.F., Hall, J.N., Hanson, A.D., 1993. CONSUME User's Guide. USDA Forest Service, Portland, OR. Gen. Tech. Rep. PNW-304.

Plucinski, M.P., Pastor, E., 2013. Criteria and methodology for evaluating aerial wildfire suppression. Int. J. Wildland Fire 22, 1144−1154.

Podur, J.J., Martell, D.L., 2007. A simulation model of the growth and suppression of large forest fires in Ontario. Int. J. Wildland Fire 16, 285−294.

Podur, J., Wotton, M., 2010. Will climate change overwhelm fire management capacity? Ecol. Model. 221, 1301−1309.

Price, O.F., Bradstock, R.A., 2012. The efficacy of fuel treatment in mitigating property loss during wildfires: Insights from analysis of the severity of the catastrophic fires in 2009 in Victoria, Australia. J. Environ. Manag. 113, 146−157.

Reinhardt, E.D., Keane, R.E., Brown, J.K., 1997. First Order Fire Effects Model: FOFEM 4.0, User's Guide. USDA Forest Service, Ogden UT. Gen. Tech. Rep. INT-GTR-344.

Reinhardt, E.D., Crookston, N.L., 2003. The Fire and Fuels Extension to the Forest Vegetation Simulator. USDA Forest Service, Ogden UT. Gen. Tech. Rep. RMRS-GTR-116.

Riley, K.L., Abatzoglou, J.T., Grenfell, I.C., Klene, A.E., Heinsch, F.A., 2013. The relationship of large fire occurrence with drought and fire danger indices in the western USA, 1984−2008: the role of temporal scale. Int. J. Wildland Fire 22, 894−909.

Roads, J., Fujioka, F., Chen, S., Burgan, R., 2005. Seasonal fire danger forecasts for the USA. Int. J. Wildland Fire 14, 1−18.

Rothermel, R.C., 1972. A Mathematical Model for Predicting Fire Spread in Wildland Fuels. USDA Forest Service, Ogden, UT. Res. Pap. INT-115.

San-Miguel-Ayanz, J., Barbosa, P., Schmuck, G., Liberta, G., Schulte, E., Viegas, D.X., 2002. Towards a coherent forest fire information system in Europe: the European forest fire information system (EFFIS). In: Viegas, D.X. (Ed.), Forest Fire Research and Wildland Fire Safety: Proceedings of IV International Conference on Forest Fire Research and 2002 Wildland Fire Safety Summit. Millpress Science Publishers.

San-Miguel-Ayanz, J., Carlson, J., Alexander, M., Tolhurst, K., Morgan, G., Sneeuwjagt, R., 2003. Current methods to assess fire danger potential. In: Chuvieco, E. (Ed.), Wildland Fire Danger Estimation and Mapping - the Role of Remote Sensing. World Scientific Publishing, Singapore, pp. 21−61.

Setzer, A.W., Sismanoglu, R.A., 2012. Risco de Fogo: Metodologia do Cálculo e Descrição sucinta da Versão 9. INPE - National Institute For Space Research, São José dos Campos, Brazil. http://queimadas.cptec.inpe.br/wrqueimadas/documentos/RiscoFogo_Sucinto.pdf (accessed 24.07.13.).

Sharples, J.J., McRae, R.H.D., Weber, R.O., Gill, A.M., 2009a. A simple index for assessing fuel moisture content. Environ. Model. Softw. 24, 637−646.

Sharples, J.J., McRae, R.H.D., Weber, R.O., Gill, A.M., 2009b. A simple index for assessing fire danger rating. Environ. Model. Softw. 24, 764−774.

Shabbar, A., Skinner, W., Flannigan, M., 2011. Prediction of seasonal forest fire severity in Canada from large-scale climate patterns. J. Appl. Meteorol. Climatol. 50, 785−799.

Show, S.B., Kotok, E.I., 1925. Fire and the forest. USDA Serv. Circular 354, 24.

Sneeuwjagt, R.J., Peet, G.B., 1985. Forest Fire Behaviour Tables for Western Australia, third ed. Department of Conservation and Land Management, Perth, WA.

Taylor, S.W., Alexander, M.E., 2006. Science, technology, and human factors in fire danger rating: the Canadian experience. Int. J. Wildland Fire 15, 121−135.

UNISDR, 2007. Hyogo Framework for Action (2005−2015): Building the Resilience of Nations and Communities to Disasters. Extract from the final report of the World Conference on Disaster Reduction (A/CONF.206/6). United Nations, Geneva, 28 pp.

VanWagner, C.E., 1987. Development and Structure of the Canadian Forest Fire Weather Index System. Can. For. Serv., For. Tech. Rep. 35, Canadian Forest Service, Ottawa, ON.

Van Wagner, C.E., Pickett, T.L., 1985. Equations and FORTRAN Program for the Canadian Forest Fire Weather Index System. Can. For. Serv., For. Tech. Rep 33, Canadian Forest Service, Ottawa, ON.

Viegas, D.X., Bovio, G., Ferreira, A., Nosenzo, A., Sol, B., 1999. Comparative study of various methods of fire danger evaluation in southern Europe. Int. J. Wildland Fire 9, 235−246.

Viney, N.R., 1992. Modelling fine fuel moisture. Ph.D. thesis, Dept. Math. Uni. New South Wales Aust. Defence Force Acad., Canberra ACT.

Won, M.S., Lee, S.Y., Lee, M.B., Ohga, S., 2010. Development and application of a forest fire danger rating system in south Korea. J. Fac. Agric. Kyushu Uni. 55, 221−229.

Wotton, B.M., 2009. Interpreting and using outputs from the Canadian forest fire danger rating system in research applications. Environ. Ecol. Stat. 16, 107−131.

Wotton, B.M., Martell, D.L., 2005. A lightning fire occurrence model for Ontario. Can. J. For. Res. 35, 1389−1401.

Wotton, B.M., Nock, C.A., Flannigan, M.D., 2010. Forest fire occurrence and climate change in Canada. Int. J. Wildland Fire 19, 253−271.

Zhdanko, V.A., 1965. Scientific basis of development of regional scales and their importance for forest fire management. In: Melekhov, I.S. (Ed.), Contemporary Problems of Forest Protection from Fire and Firefighting. Lesnaya Promyshlennost' Publisher, Moscow, pp. 53−86 (in Russian).

Zommers, Z., Singh, A. (Eds.), 2014. Reducing Disaster: Early Warning Systems for Climate Change. Springer, New York, p. 387.

Chapter 12

Postfire Ecosystem Restoration

V. Ramon Vallejo
CEAM, Parque Tecnológico, Ch. Darwin 14, Paterna, Spain; Dept. Biologia Vegetal, Universitat de Barcelona. Diagonal 643, Barcelona, Spain

J. Antonio Alloza
CEAM, Parque Tecnológico, Ch. Darwin 14, Paterna, Spain

ABSTRACT

Postfire restoration is meant to mitigate or reverse negative fire impacts. Impacts are related to fire regime and its interactions with ecosystem fire resilience. In the case of severe fire regimes, the main ecological impacts affect nutrient budget, soil-erosion risk, and the reduction of biodiversity. Planning postfire restoration requires the identification of the specific degradation processes triggered by fire, including their time and spatial dimensions, and vulnerable ecosystems. Restoration should address identified vulnerable areas, and mitigate soil erosion and runoff risk in the short term, and the recovery of nutrient cycling and keystone plant species in the longer term. We present the approach developed for assessing postfire restoration in the Mediterranean basin based on the prediction of soil-erosion risk and vegetation vulnerability.

12.1 INTRODUCTION

Most of the regions supporting enough productivity are affected by wildfires nowadays (Pausas and Ribeiro, 2013). Large wildfires produce impacts both at the global and at the local scale. Globally, through atmospheric emissions of greenhouse gases and particles (e.g., Levine et al., 1999; Page et al., 2002; Randerson et al., 2006; Bowman et al., 2009). Locally through the impacts on air quality and through the direct and indirect impact on the ecosystems and landscapes, including human structures. Restoring resilient landscapes will reduce damages and suppression costs in the long term (Ryan and Opperman, 2013).

According to fire severity, fire produces direct damage to vegetation (e.g., Lloret and Zedler, 2009) and fauna, and direct and indirect impacts on soils through heat release and ash deposition, and early postfire degradation. Impacts usually affect nutrient availability, biological activity and, especially for high ground fire severity, soil physical properties (Figure 12.1). In ground high-intensity smouldering fires affecting peaty soils, combustion not only release

Wildfire Hazards, Risks, and Disasters. http://dx.doi.org/10.1016/B978-0-12-410434-1.00012-9

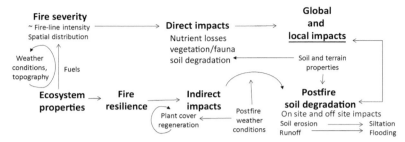

FIGURE 12.1   Main fire impacts and related controlling factors.

huge amounts of carbon dioxide (CO_2) but it also consumes and deeply modifies the soil resource, producing long-term damages (Rein et al., 2008). Fire produces loss of organic matter and nutrients by volatilization, especially nitrogen (N) and sulfur (S), phosphorus (P) to a lesser extent, and smoke particles according to fire intensity (Raison et al., 1985a,b). In spite of net nutrient ecosystem losses, in most of the cases, the soil is enriched by nutrients in the ashes right after the fire. Ashes are usually poor in mineral nitrogen and rich in P, calcium (Ca), and potassium (K), more so in high-intensity fires. During the first postfire months, there is a peak of N mineralization, especially nitrification (Serrasolses and Vallejo, 1999). Overall, the burned ecosystem is often rich in available nutrients in the short term, both for microorganisms and for the initial regenerating plants (Walker et al., 1986; Attiwill and Leeper, 1987; Serrasolsas and Khanna, 1995). However, soil microbial biomass and diversity may take a long time to recover to prefire levels (Mabuhany et al., 2006).

Nutrient losses in recurrent and high-intensity fires could significantly reduce ecosystem nutrient budget and soil fertility in extremely poor soils, and this could be especially relevant for P (Raison et al., 2009) in highly weathered, old soils (Specht and Moll, 1983; Lamont, 1995). Repeated fires may deplete labile N pools in soil and litter (Raison et al., 1993a), and reduce plant productivity, even for plant species having a large pool of below-ground carbohydrates and nutrient reserves (Ferran et al., 2005).

Soil physical properties are only affected by fire under high intensity on the soil surface. Two effects are particularly relevant for ecosystem functioning, both reducing soil water infiltration capacity: (1) direct formation of a hydrophobic layer (Doerr et al., 2009), especially in sandy soils; and (2) indirect formation of a physical soil crust after the first postfire rain events, especially in poorly structured, silty soils (Bautista et al., 1996; Llovet and Vallejo, 2010). Reduced soil infiltration capacity together with the transient lack/reduced plant and forest floor cover right after fire facilitates high postfire erosion and runoff risk (Vallejo and Alloza, 1998; Scott and Curran, 2009). This is especially acute in regions where the fire season is followed by the rainy season and more so for regions with frequent heavy rains such as the Mediterranean (Figure 12.2).

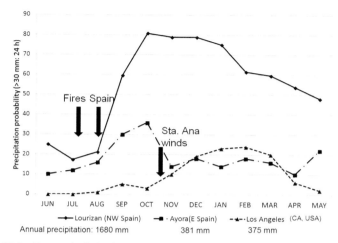

FIGURE 12.2   Temporal relationships between fire and heavy rain occurrence in three contrasted examples from Spain and California. Rain erosivity is indicated by the probability of having a rain event with a higher intensity than 30 mm/24 h, as surrogate. In the two Spanish sites, the main fire season (summer) is followed by the peak of rain intensity (autumn), both for a representative dry Mediterranean site (Ayora) and for an Atlantic temperate site (Lourizan). In the Mediterranean site of California (Los Angeles), for annual precipitation similar to that in Ayora, the fire season often occurs in autumn, in relation to the Santa Ana winds, and the rainy season occurs in winter. In the three contrasted situations, fire occurrence is followed by high risk of heavy rains, hence of postfire soil erosion.

Severe erosion may produce deep soil quality degradation (Shakesby, 2011) that may lead to irreversible losses of topsoil particles at the ecological time frame, together with nutrients and seeds. Postfire soil degradation may also produce downstream damages (flooding, siltation) to ecosystems and human structures. Therefore, among the various impacts of wildfires in the ecosystem discussed above, soil erosion could be the most irreversible, beyond restoration capabilities as far as soil formation is an extremely slow natural process, and artificial production of soil-like material is unaffordable for large areas. Hence, the mitigation of postfire soil erosion should be a first priority in postfire management (Vallejo and Alloza, 1998).

12.2 DO WE NEED TO MANAGE ECOSYSTEM RECOVERY AFTER WILDFIRES?

Large fires have strong social impacts. Often, social organizations and the media call for immediate actions, and the problem becomes a political issue, at least in the short term. Restoration projects are extremely expensive, both in economic and in energy terms. Large fires, especially megafires, affect large areas often beyond the logistic capacity of agencies to materially address restoration actions for the whole burned areas. Therefore, postfire restoration projects should be carefully prioritized, and scientifically and technically justified.

Postfire restoration assessment should be based on the understanding of how fire regime is affecting ecosystem fire resilience. The rationale of postfire restoration should respond to the following chained sequence of questions: why, what, when, and how—restoring burned areas.

12.2.1 Why? Assessment of Postfire Restoration Needs

12.2.1.1 Setting Restoration Objectives

Restoration is always local, even when considering a global framework. Therefore, specific restoration objectives could be very diverse (Vallejo and Alloza, 2012) according to biophysical and socioeconomic conditions (Rojo et al., 2012). However, we could establish a common background for the scope of this paper assuming a global objective of avoiding postfire damages onsite and offsite on ecosystem and structures, and ecosystem conservation and/or improvement, taking the baseline reference of prefire ecosystem. From this minimum, any better structural and functional ecosystem could be used as reference in the site (Aronson and Vallejo, 2006).

Assessing postfire impacts and fire ecosystem resilience should provide the scientific basis for assessing postfire restoration needs. This is the critical step in postfire restoration assessment. The challenge is to find out what combination of fire severity and ecosystem properties, related to their fire resilience, makes the ecosystem vulnerable so to deserve postfire restoration actions. This ecological perspective should be complemented with the potential human impacts and perspectives. Summarizing, ecological fire impacts include: Nutrient losses; temporary loss of plant cover and fauna (the question is how long and how reversible?); and soil degradation risk and modification of water regime, including their effects downslope/downstream. The factors affecting fire impact are fire regime and also ecosystem and site characteristics (fuels, topography, and weather conditions) (Figure 12.3).

12.2.2 What Is Being Restored?

12.2.2.1 Restoring Soil Productivity

The risk of postfire soil erosion is related to the eventual direct degradation of the soil surface by fire and, usually more relevant, the temporary lack of protective plant and forest floor cover. Therefore, plant cover recovery rate is the critical factor in controlling postfire soil erosion risk (Ferran et al., 1992; Vallejo et al., 1999). Among the factors affecting soil erosion, plant and/or litter cover is probably the most feasible to be modified through management. The recovery of the physical, chemical, and biological soil properties are much dependent on postfire soil organic matter dynamics, which is also driven by

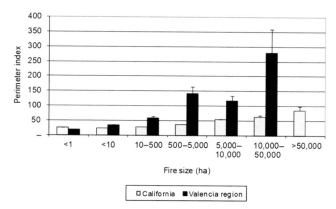

FIGURE 12.3 Variation of the fire perimeter index for different-size classes in two Mediterranean regions: California and Valencia Region (eastern Spain). Perimeter index for burned areas: perimeter^2/surface. The perimeter index increases with fire size, more in the Valencia Region where landscapes are highly heterogeneous in terms of different intermingled land uses. Probably, for the same reason, very large fires (>50,000 ha) do not occur in the Valencia Region. *Elaborated from cartographic data from the California Department of Forestry and Fire Protection (Fire and Resource Assessment Program) and Conselleria Gobernación y Justicia (Generalitar Valenciana).*

postfire vegetation regeneration. The rate of postfire regeneration is essentially dependent on the postfire regeneration strategies of dominant plants (Keeley et al., 2012). In terms of carbon (C) and nutrient cycles and soil biological activity recovery, the driver is the recovery of litter inputs to the soils, both in terms of quantity and quality (Ferran and Vallejo, 1992; Kaye et al., 2010), and the organic matter decomposition rate, related to the biophysical soil environment. Postfire nutrient availability dynamics would be controlled by the short-term direct and indirect effects of fire in a first phase (from days to few months), and soon would respond to feedbacks with vegetation regeneration.

Legumes are known to start fixing N in the early stages of development after fire (Casals et al., 2005) and would eventually compensate net N losses produced by burning. Of course, the time required for recovering the prefire N status would depend on the magnitude of N losses (fire severity, postfire leaching, and erosion), fire frequency, and the abundance and fixing activity of legumes. Raison et al. (2009) stressed that with legumes N-fixation may be limited by soil P availability in unfertile soils. In these soils, P losses by combustion could be the most critical for soil fertility and ecosystem conservation (Raison et al., 2009) as far as new natural P inputs are negligible in highly weathered soils. Inorganic P addition would significantly enhance N and P fertility recovery in low fertility soils (Raison et al., 1993b).

In summary, soil conservation and recovery after burning would require first controlling soil erosion and runoff (Robichaud, 2009). In a second phase, vegetation recovery would be the key factor for the recovery of soil functions and structure. In extreme cases of naturally very low fertility soils, artificial soil P amendment could be very efficient and cost effective.

Maintaining long-term ecosystem productivity will also require maintaining long-term nutrient cycling, and this may be especially an issue in postfire management for P-poor soils (Raison et al., 2009), and for very severe or high recurrence fires. The use of biosolids increases soil organic matter, soil fertility, and introduced seedling growth in reforestation projects (Valdecantos et al., 2011). However, care should be taken to not overfertilize burned ecosystems (Fuentes et al., 2010), leading to soil eutrophication and the proliferation of weeds.

12.2.2.2 The Recuperation of Prefire Vegetation

In fire-prone areas, plant species show several adaptations to survive fires. In these regions, there are also species not surviving canopy fires, although not many (Lloret and Zedler, 2009). As a working hypothesis, we could assume that no species is adapted to all fire regimes.

Changes in fire regime may trigger switches between vegetation in savannas, boreal forests, and temperate forests (USA and Australia) (Adams, 2013). For example, highly severe fires may produce the substitution of conifers by hardwoods in boreal forests (Cai et al., 2013; Girardin et al., 2013). The ability of plant species populations to survive fire, their recovery mode, and rate (Lloret and Zedler, 2009), as well as the rates of fuel production and their characteristics that may facilitate new fires in the short term (Baeza et al., 2011), differ for different vegetation types.

In fire-prone, fire-dependent regions, most plant species have developed mechanisms to efficiently recover after wildfires, some at the individual level, and other at the population level (e.g., Lloret and Zedler, 2009; Keeley et al., 2012). The ability of the dominant species to regenerate after fire is an essential factor in the fire resilience of the ecosystem. For high-severity canopy fires, some species may survive at the individual level by vegetative regrowth from below-ground or protected above-ground buds, the resprouter species (obligate resprouters, Keeley et al., 2012). Resprouting is widely distributed in the world, across taxonomic groups and biomes (Lloret and Zedler, 2009). The degree of resprouting is very variable between species (Reyes and Casal, 2008), plant age, and fire regime (Lloret and Zedler, 2009). Efficient resprouting keeps the same species composition and abundance in the plant community soon after fire (Ferran et al., 1992). The second major postfire regeneration strategy is seeding recruitment (obligate seeders if they lack resprouting capacity, Keeley et al., 2012), mostly from the soil seed bank and also from the canopy seed bank in highly adapted species. Soil and canopy seed banks are dependent on fire intensity, frequency, and interval. High fire intensity at the soil surface may kill the most superficial part of the seed bank, and the most sensitive species, whereas it may stimulate seed germination for adapted species below a lethal intensity (maximum temperature and heating duration) threshold. High fire frequency could eventually deplete the soil seed bank, and a short fire interval could eradicate species dependent on the buildup of a sufficient canopy seed bank, when the interval is shorter than the time required to reach sexual maturity (immaturity risk, e.g., serotinous pine species, Pausas et al., 2004). Facultative seeders are those species that may regenerate both by resprouting and by seedling recruitment (Keeley et al., 2012). Seedling recruitment is more dependent on the soil moisture and temperature conditions after fire than resprouting, and it is generally slower in the recovery of plant cover (Ferran et al., 1992; Vallejo and Alloza, 1998). The type of vegetation regenerating would depend on the available seed and surviving bud bank in the site, and that is essentially related to prefire vegetation composition. Plant individuals resprouting or germinating after fire would respond in their growth rate and competition capacity according to their ability to thrive in the postfire soil habitat.

According to Shlisky et al. (2007), ecoregions could be classified at the global scale in fire dependent, fire sensitive, and fire independent, according to the relationships of their ecosystems with fire regime characteristics. In fire-dependent ecosystems, most of the species have evolved with fires, and fire is considered an evolutionary factor (Pausas and Keeley, 2009). The opposite happens in fire-sensitive ecosystems, where fire is mostly introduced by recent human activities. Fire-independent ecosystems do not sustain fire propagation because of insufficient productivity (i.e., fuel load and continuity). Therefore, the impact of wildfires would depend on the evolutionary, fire-related traits of their dominant plants, and also on the degree of human transformation/degradation that may modify ecosystem sensitivity to fire. In principle, we could expect strong fire impacts in fire-sensitive ecosystems. At a global scale, we could consider some representative situations:

Fire-dependent, fire-prone regions include: Little altered ecosystems (e.g., boreal); deeply altered ecosystems and landscapes (geomorphology, soils, species composition, e.g., Mediterranean basin); and structurally altered ecosystems (by invasive species, and/or changes in forest structure and/or fire regime) that change from fire resistant to fire vulnerable, for example, temperate and subalpine coniferous forests.

Fire-sensitive regions: In these regions fire is induced by forest exploitation (e.g., tropical rain forests, Cochrane, 2003), and climate change and increased human ignitions (e.g., newly affected mountain forests in the Mediterranean basin, Ganatsas et al., 2012).

In naturally fire-free, fire-sensitive regions, human activities may have introduce fire for land clearing, forest exploitation (e.g., opening tropical rain forests changing fuels and microclimate), or increasing tourism pressure leading to negligence ignitions (high elevation forests in the Mediterranean). Climate change is an additional fire-regime modifier (Krawchuk et al., 2009).

In fire-prone regions, under historical fire incidence, the question is if any existing fire regime may drive to irreversible ecosystem degradation. This situation could appear after a new combination of ecosystem properties and/or on fire regime, for example, abrupt changes in fuel characteristics, and/or in the ignition pattern, and/or the weather conditions. Some documented examples related to direct or indirect effects of human activities:

- Human ignitions highly increasing fire recurrence for improving pastures, arson, and negligence in highly populated areas (Catry et al., 2010).
- Indirect modification of fuels through land use history effects on plant succession, and nonfire-related ecosystem degradation. For example, the development of fire-prone shrublands in old fields in the Mediterranean (Baeza et al., 2007; Duguy and Vallejo, 2008).
- Fire suppression effects on fuels and fire regime (e.g., Minnich, 2001; Miller et al., 2012).
- Invasive plant species modifying fuels and fire regime (Keeley, 2006).
- Wasteful forest exploitation dramatically producing a large accumulation of slash and leading to extremely severe wildfires (e.g., large fires in North America during the late nineteenth to early twentieth centuries, Williams et al., 2013).

All the above factors would be favored by climate change in many regions of the world, although they could be limited in others (Krawchuk et al., 2009).

12.2.3 When and How? Postfire Restoration Approaches

Ecological impacts of forest fires may be very diverse according to fire behavior and terrain and ecosystem properties (Figure 12.1); therefore, caution should be taken in making broad generalizations about fire impacts (Raison et al., 2009). However, postfire restoration may have some common grounds of wide applicability: the more common these are, the more narrowly defined the specific restoration objectives will be. Short-term postfire restoration aiming at minimizing impacts, and at ecosystem stabilization (Robichaud, 2009), may have general principles of wide applicability (e.g., the USDA BAER catalog of emergency rehabilitation treatments, Napper, 2006). On the contrary, long-term restoration objectives may be extremely diverse depending to the variety of social–ecological systems (Vallejo and Alloza, 2012), thus making it difficult to develop general guidelines. Long-term restoration would aim at restoring ecosystem integrity (function and structure, including biodiversity) together with ecosystem services.

12.3 THE CASE OF MEGAFIRES

Megafires are increasing all over the world and accounting for a large part of fire impacts (Adams, 2013). Hence, megafires deserve special attention in both fire management and postfire restoration. Although the concept of megafires has no precise physical definition, from an operational point of view they are considered those that produce high impacts in both ecosystems and society (Williams et al., 2011). From the perspective of postfire restoration, the question is: "What makes megafires different?"

Megafires affect large areas producing high fire line intensity that should result in a global high fire severity. Therefore, from the perspective of postfire restoration, two distinctive features could be critical: large burned areas and high global fire severity. Megafires produce high global impacts through the emissions of greenhouse gases and particles to the atmosphere. However, ecosystem conservation and restoration deal mostly with local impacts of fires on ecosystems in the landscape. Global impacts can be more or less linearly related to fire size, whereas local impacts cannot.

12.3.1 The Role of Fire Size

The question is: "To what extent does fire size matter to ecosystem recovery?". Large fires are characterized by their high intensity, and thus usually high severity as well. From the perspective of postfire recovery, early vegetation recovery does not depend on fire size as most short-term regeneration relies on endogenous, local propagules (local seed bank and/or resprouts). Fire size is very relevant for sensitive species that are eradicated by canopy fires; these species have to recolonize the site from outside of the fire perimeter, or from unburned or low-severity burned patches within the perimeter. In this case, dissemination distance and the implicated dissemination vectors are critical. Therefore, recolonization may be slow for very large fires.

The form of the fire perimeter is also relevant. Figure 12.4 shows that the proportion of perimeter contact between burned/unburned areas increases with the size of the burned area. This should be related to the increase of the irregularity of the perimeter as fire size increases, and this may facilitate the postfire colonization of plants and animals from unburned, neighboring populations. A second factor to consider is fire size in relation to the ecosystem or species distribution area and its connectivity to unaffected areas, for example, relatively small fires could affect the total area naturally accessible for a species' propagule recolonization. This is especially evident in islands. Fire size and shape are related to land use structure in the landscape. Extremely large fires can only occur for large continuous areas of fuels, that is, for areas with a relatively low human occupation of the territory. In highly human-disturbed regions such as the Mediterranean basin, fire size is very often constrained by fuel continuity.

12.3.2 The Role of Fire Severity in the Landscape

Fire severity is often considered as a critical factor in fire impacts and postfire regeneration. For the purpose of this chapter, we will use the often-fuzzy

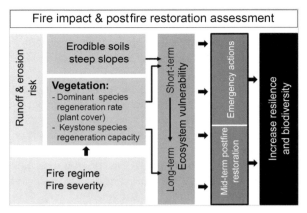

FIGURE 12.4 Outline of the fire impact and postfire restoration assessment protocol elaborated for Mediterranean-basin conditions. See the text for details.

concept of fire severity in the sense of Keeley (2009), that is, organic matter loss produced by fire intensity. In principle, high fire severity is related to higher plant mortality and lower postfire regeneration of extant plants, both resprouts and seed-bank recruitment. This holds true for plant mortality, by definition of severity, but this is not always the case for postfire regeneration of canopy burned plants (Lloret and Zedler, 2009). Pausas et al. (2003) showed that the fire-adapted Aleppo pine had better postfire growth for high-severity crown fires. We could assume that some species are well adapted to high-severity canopy fires.

Very often, fire severity categories are used in highly relative terms (Hartford and Frandsen, 1992), that is, the degree of consumption of litter and the proportion of plants killed by fire. This makes comparison of the impacts of severity among ecosystems difficult. The range of possible fire severities and their specific location in the ecosystem are very much controlled and limited by the accumulation, characteristics, and spatial distribution of fuels within the ecosystem, both vertically and horizontally. This is a feature associated with ecosystem type and management. Fuel distribution in the ecosystem depth constraints the possible amounts of heat release, maximum temperatures, and residence times in each ecosystem (fuel) compartment. For example, Mediterranean forests seldom show high severity at the soil surface, even for high-intensity crown fires, because of the poor litter load, and this might be extremely different in a boreal forest, even more over a peaty soil. Therefore, the role of fire severity in ecosystem regeneration, as such general approach, is difficult to be properly assessed when comparing different ecosystem types, and often gives way to overgeneralizations.

Fire severity is very often highly heterogeneous at all spatial scales, at the landscape and even at the patch scale (Turner et al., 1994; Neary et al., 1999). In most of the cases, we could expect that the larger the fire, the higher the area affected by high severity. The spatial heterogeneity in fire severity, including unburned patches within the fire perimeter, could be very important for the recolonization of the affected area (Schoennagel et al., 2008). Unburned and low-severity patches scattered in the burned landscape would provide nearer seed sources for recolonization.

From the restoration practice perspective, extremely large restoration projects are not feasible. Therefore, any restoration that might be envisaged would have to focus on small hot spots and consider a very large time framework. Large fires produce a high diversity of impacts in the landscape. For this reason, ecosystem recovery is expected to be spatially diverse at various scales, as is the necessity of postfire management actions. The challenge is to identify the patches that require restoration actions and determine how to prioritize them. Long-term restoration should take into account the landscape dimension, for example, natural colonization potential in relation to connectivity (Stephens et al., 2010).

12.4 A MEDITERRANEAN-BASIN APPROACH

In this section, we summarize the 20-year postfire restoration experience in the Mediterranean basin (Vallejo and Alloza, 1998; Alloza and Vallejo, 2006; Vallejo et al., 2009; Vallejo and Alloza, 2012) that has been developed in eastern Spain and tested in most of the Euro-Mediterranean countries (Moreira and Vallejo, 2009; Vallejo et al., 2012a). From this research background, we have developed a protocol for assessing ecological fire impacts and postfire restoration actions (Figure 12.4; Alloza et al., 2014).

In synthesis, the approach includes (1) Establishment of specific management objectives for the burned areas. Stakeholders' participatory approaches are recommended for the development of restoration objectives (Rojo et al., 2012). The assumed objectives of general application are avoid or minimize short-term damages (erosion, flash floods) and stabilize the soil; increase fire resilience and biodiversity; improve fire prevention in the perspective of new fires. (2) Identification of fire-vulnerable ecosystems. This includes the prediction of runoff and soil-erosion risk and the prediction of dominant plant species regeneration capability (resistance/resilience, regeneration rate) as a function of fire severity and of landscape and ecosystem characteristics. For Mediterranean-basin ecosystems, the abundance of resprouters is key in explaining plant recovery rate and ecosystem resilience (Vallejo and Alloza, 1998). The assessment is based on the use of relevant maps and geographic information system, and field surveys conducted immediately after fire. On the basis of this assessment: (3) Timely application of specific techniques to mitigate degradation and assist regeneration. Different steps are considered according to the timing of postfire risks (Figure 12.4, Duguy et al., 2012; Vallejo and Alloza, 2012): (a) short-term, less than one year, for emergency measures to mitigate damages and stabilize the soils (see the exhaustive BAER catalog for emergency rehabilitation techniques, Napper, 2006); (b) longer-term (2—5 years) for assisting natural regeneration and improving ecosystem biodiversity and resilience. The introduction of native woody resprouters is recommended to increase fire resilience and reduce fire hazard in degraded sites. Plantation techniques were especially developed for improving water limitations for outplanted seedlings (Vallejo et al., 2012b). Monitoring and evaluation should be implemented in all restoration actions (Bautista et al., 2010) in the framework of adaptive management to address the inherent uncertainties of ecological restoration.

The approach is, of course, conditioned by the biophysical and social contexts in which it was developed. The main relevant features of the Euro-Mediterranean domain with respect to fire impacts and postfire restoration are:

- Fire-dependent (according to the scheme proposed by Shlisky et al. (2007); see above), fire-prone region (Keeley et al., 2012): The Mediterranean is the most vulnerable to global change of the European regions, and increased risk of forest fires is one of the critical potential impacts (Schröter et al., 2005).
- Crown fires could be severe, for both trees and shrub canopies, the latter both in the forest understory and in shrublands.
- Generally, low fire severity affects the forest floor and topsoil due to the relatively low amount of forest floor fuel. Of course, exceptions may occur for fuel models with forest slash after logging.
- Abundant woody sclerophyllous plant species, extremely vigorous resprouters, surviving even for high-intensity and recurrent fires (Trabaud, 1990; Delitti et al., 2005).
- Contrasted recovery rate between woody obligate seeders and obligate resprouters (Vallejo and Alloza, 1998). Shrubby obligate seeders are often opportunistic pioneer species, for example, in old fields, and accumulate risky fine and dead fuel in the short term (Baeza et al., 2011).
- Low abundance of invasive plant species potentially modifying the fire regime in the Mediterranean basin (so far) (Vallejo et al., 2012a), very much in contrast with other Mediterranean regions in the world such as California (Keeley et al., 2012).
- High rain erosivity right after (autumn) the fire season (summer) (Figure 12.2.).
- Long-term human-induced ecosystem transformation and degradation (Vallejo and Alloza, 1998) by intensive and extensive land over-exploitation, including fire (Keeley et al., 2012).
- Recent extensive land abandonment, dramatically increasing fire hazard (Pausas and Vallejo, 1999).
- Wildfires often affect the wildland—urban interface and periurban areas. This generates great population security risks, social alarm, and demand for civil protection measures.

- Forest management and postfire restoration strongly dependent on forest administrations and on public funds.

How applicable is this experience to other ecosystems types, within (e.g., other Mediterranean type ecosystems, such as California or South Africa) and outside the same bioregion, is still an open question.

12.5 CONCLUSION

Postfire restoration should be based on the assessment of fire impacts in relation to ecosystem resilience. Fire impacts are very diverse for different regions and fire regimes. Therefore, postfire restoration approaches should be also diverse and respond to the local, specific conditions addressed. However, common methodological approaches could be possible on the basis of the prediction of ecosystems response to the various fire regimes associated to these ecosystems. The interactions of fire regime with vegetation postfire regeneration strategy are probably essential elements to better predict fire impacts and prioritize restoration actions. Restoration actions and techniques should be tuned to the specific ecological and social risks associated with direct and indirect fire impacts, both considering the spatial distribution of risks and the timing of degradation process. The actions should be framed by the long-term management objectives defined for the burned areas.

ACKNOWLEDGMENTS

The preparation of this review was financed by the Generalitat Valenciana, and the projects GRACCIE (Ministerio de Ciencia e Innovación, Programa Consolider-Ingenio 2010, CSD2007-00067), PROMETEO/2009/006, and FUME (European Commission, FP7, grant agreement 243888).

REFERENCES

Adams, M.A., 2013. Mega-fires, tipping points and ecosystem services: managing forests and woodlands in an uncertain future. Forest Ecol. Manage. 294, 250–261.

Alloza, J.A., Garcia, S., Gimeno, T., Baeza, J., Vallejo, V.R., Rojo, L., Martinez, A., 2014. Guía técnica para la gestión de montes quemados. Protocolos de actuación para la restauración de zonas quemadas con riesgo de desertificación. Ministerio de Agricultura, Alimentación y Medio Ambiente. Madrid.

Alloza, J.A., Vallejo, V.R., 2006. Restoration of burned areas in forest management plans. In: Kepner, W.G., Rubio, J.L., Mouat, D.A., Pedrazzini, F. (Eds.), Desertification in the Mediterranean Region: A Security Issue. Springer, pp. 475–488.

Aronson, J., Vallejo, V.R., 2006. Challenges for the practice of ecological restoration. In: van Andel, J., Aronson, J. (Eds.), Restoration Ecology. Blackwell Science, Oxford, UK, pp. 234–247.

Attiwill, P.M., Leeper, G.W., 1987. Forest Soils and Nutrient Cycles. Melbourne University Press, Melbourne, Australia.

Baeza, M.J., Santana, V.M., Pausas, J.G., Vallejo, V.R., 2011. Successional trends in standing dead biomass in Mediterranean basin species. J. Veg. Sci. 22, 467–474.

Baeza, J., Valdecantos, A., Alloza, J.A., Vallejo, V.R., 2007. Human disturbance and environmental factors as drivers of long-term post-fire regeneration patterns in Mediterranean forests. J. Veg. Sci. 18, 243–252.

Bautista, S., Bellot, J., Vallejo, V.R., 1996. Mulching treatment for post-fire soil conservation in a semiarid ecosystem. Arid Soil Res. Rehabil. 10, 235–242.

Bautista, S., Orr, B.J., Alloza, J.A., Vallejo, V.R., 2010. Evaluating the restoration of dryland ecosystems in the northern Mediterranean. In: Scheneier-Madanes, G., Courel, M.F. (Eds.), Water and Sustainability in Arid Regions. Springer, New York, NY, USA, pp. 295–310.

Bowman, D., Balch, J.K., Artaxo, P., Bond, W.J., Carlson, J.M., Cochrane, M.A., D'Antonio, C.M., DeFries, R.S., Doyle, J.C., Harrison, S.P., Johnston, F.H., Keeley, J.E., Krawchuk, M.A., Kull, C.A., Marston, J.B., Moritz, M.A., Prentice, I.C., Roos, C.I., Scott, A.C., Swetnam, T.W., van der Werf, G.R., Pyne, S.J., 2009. Fire in the Earth system. Science 324, 481–484.

Cai, W., Yang, J., Liu, Z., Hu, Y., Weisberg, P.J., 2013. Post-fire tree recruitment of a boreal larch forest in Northeast China. Forest Ecol. Manage. 307, 20–29.

Casals, P., Romanyà, J., Vallejo, V.R., 2005. Short-term nitrogen fixation by legume seedlings and resprouts after fire in Mediterranean old-fields. Biogeochemistry 76, 477–501.

Catry, F., Rego, F., Silva, J., Moreira, F., Camia, A., Ricotta, C., Conedera, M., 2010. Fire starts and human activities. In: Sande Silva, J., Rego, J.F., Fernandes, P., Rigolot, E. (Eds.), Towards Integrated Fire Management—Outcomes of the European Project Fire Paradox. European Forest Institute Research Report 23, Joensuu, Finland, pp. 9–22.

Cochrane, M.A., 2003. Fire science for rainforests. Nature 421, 913–919.

Delitti, W., Ferran, A., Trabaud, L., Vallejo, V.R., 2005. Effects of fire recurrence in *Quercus coccifera* L. shrublands of the Valencia region (Spain): I. Plant composition and productivity. Plant Ecol. 177, 57–70.

Doerr, S.H., Shakesby, R.A., MacDonald, L.H., 2009. Soil water repellency: a key factor in post-fire soil erosion. In: Cerdá, A., Robichaud, P.R. (Eds.), Fire Effects on Soils and Restoration Strategies. Science Publishers, Enfield, NH, USA, pp. 197–224.

Duguy, B., Alloza, J.A., Baeza, M.J., De la Riba, J., Echeverría, M.T., Ibarra, P., Llovet, J., Pérez-Cabello, F., Rovira, P., Vallejo, V.R., 2012. Modelling the ecological vulnerability to forest fires in Mediterranean ecosystems using geographic information technologies. Environ. Manage. 50, 1012–1026.

Duguy, B., Vallejo, V.R., 2008. Land-use and fire history effects on post-fire vegetation dynamics in eastern Spain. J. Veg. Sci. 19, 97–108.

Ferran, A., Vallejo, V.R., 1992. Litter dynamics in post-fire successional forests of *Quercus ilex*. Vegetation 99–100, 239–246.

Ferran, A., Serrasolsas, I., Vallejo, V.R., 1992. Soil evolution after fire in *Quercus ilex* and *Pinus halepensis* forests. In: Teller, A., Mathy, P., Jeffers, J.N.R. (Eds.), Responses of Forest Ecosystems to Environmental Changes. Elsevier, Amsterdam, pp. 397–405.

Ferran, A., Delitti, W., Vallejo, V.R., 2005. Effects of fire recurrence in *Quercus coccifera* L. shrublands of the Valencia Region (Spain): II. Plant and soil nutrients. Plant Ecol. 177, 71–83.

Fuentes, D., Valdecantos, A., Llovet, J., Cortina, J., Vallejo, V.R., 2010. Fine-tuning of sewage sludge application to promote the establishment of *Pinus halepensis* seedling. Ecol. Eng. 36, 1213–1221.

Ganatsas, P., Daskalakou, E., Paitaridou, D., 2012. First results on early post-fire succession in an *Abies cephalonica* forest (Parnitha National Park, Greece). iForest 5, 6–12.

Girardin, M.P., Ali, A.A., Carcaillet, C., Gauthier, S., Hély, C., Le Goff, H., Terrier, A., Bergeron, Y., 2013. Fire in managed forests of eastern Canada: risks and options. Forest Ecol. Manage. 294, 238–249.

Hartford, R.A., Frandsen, W.H., 1992. When it's hot, it's hot – or maybe it's not (surface flaming may not portend extensive soil heating). Int. J. Wildland Fire 2, 139–144.

Kaye, J.P., Romanyà, J., Vallejo, V.R., 2010. Plant and soil carbon accumulation following fire in Mediterranean woodlands in Spain. Oecologia 164, 533–543.

Keeley, J.E., 2006. Fire management impacts of invasive plants in the western United States. Conserv. Biol. 20, 375–384.

Keeley, J.E., 2009. Fire intensity, fire severity and burn severity: a brief review and suggested usage. Int. J. Wildland Fire 18, 116–126.

Keeley, J.E., Bond, W.J., Bradstock, R.A., Pausas, J.G., Rundel, P.W. (Eds.), 2012. Fire in Mediterranean Ecosystems. Ecology, Evolution and Management. Cambridge University Press, New York, NY, USA.

Krawchuk, M.A., Moritz, M.A., Parisien, M.-A., Van Dorn, J., Hayhoe, K., 2009. Global pyrogeography: the current and future distribution of wildfire. PLoS ONE 4 (4), e5102. http://dx.doi.org/10.1371/journal.pone.0005102.

Lamont, B.B., 1995. Mineral nutrient relations in Mediterranean regions of California, Chile, and Australia. In: Arroyo, M.T.K., Zedler, P.H., Fox, M.D. (Eds.), Ecology and Biogeography of Mediterranean Ecosystems in Chile, California, and Australia. Springer-Verlag, Berlin, pp. 211–235.

Levine, J.S., Bobbe, T., Ray, N., Singh, A., Witt, R.G. (Eds.), 1999. Wildland Fires and the Environment: A Global Synthesis. UNEP/DEIAEW/TR.99-1, Sioux Fall, SD, USA.

Lloret, F., Zedler, P.H., 2009. The effects of fire on vegetation. In: Cerdá, A., Robichaud, P.R. (Eds.), Fire Effects on Soils and Restoration Strategies. Science Publishers, Enfield, NH, USA, pp. 257–295.

Llovet, J., Vallejo, V.R., 2010. Post-fire dynamics of soil surface compaction in relation to the previous stage of land abandonment. Cuaternario y Geomorfología 24 (3–4), 53–62.

Mabuhany, J.A., Isagi, Y., Nakagoshi, N., 2006. Wildfire effects on microbial biomass and diversity in pine forests at three topographic positions. Ecol. Res. 21, 54–63.

Miller, J.D., Collins, B.M., Lutz, J.A., Stephens, S.L., van Wagtendonk, J.W., Yasuda, D.A., 2012. Differences in wildfires among ecoregions and land management agencies in the Sierra Nevada region, California, USA. Ecosphere 3 (9), 80. http://dx.doi.org/10.1890/ES12–00158.1.

Minnich, R.A., 2001. An integrated model of two fire regimes. Conserv. Biol. 15, 1549–1553.

Moreira, F., Vallejo, V.R., 2009. What to do after fire? Post-fire restoration. In: Birot, Y. (Ed.), Living with Fires. European Forest Institute Discussion Paper 15. EFI, Joensuu, Finland, pp. 53–58.

Napper, C., 2006. BAER—Burned Area Emergency Response Treatments Catalog. USDA Forest Service, San Dimas, CA, USA.

Neary, D.G., Klopatek, C.C., DeBano, L.F., Ffolliott, P.F., 1999. Fire effects on belowground sustainability: a review and synthesis. Forest Ecol. Manage. 122, 51–71.

Page, S.E., Siegert, F., Rieley, J.O., Boehm, H.-D.V., Jaya, A., Limin, S., 2002. The amount of carbon released from peat and forest fires in Indonesia during 1997. Nature 40, 61–65.

Pausas, J.G., Bladé, C., Valdecantos, A., Seva, J.P., Fuentes, D., Alloza, J.A., Vilagrosa, A., Bautista, S., Cortina, J., Vallejo, V.R., 2004. Pines and oaks in the restoration of Mediterranean landscapes of Spain: new perspectives for an old practice – a review. Plant Ecol. 171, 209–220.

Pausas, J.G., Keeley, J.E., 2009. A burning story: the role of fire in the history of life. BioSience 59, 593−601.

Pausas, J.G., Ouadah, N., Ferran, A., Gimeno, T., Vallejo, V.R., 2003. Fire severity and seedling establishment in *Pinus halepensis* woodlands, eastern Iberian Peninsula. Plant Ecol. 169, 205−213.

Pausas, J.G., Ribeiro, E., 2013. The global fire-productivity relationship. Glob. Ecol. Biogeogr. 22, 728−736.

Pausas, J.G., Vallejo, V.R., 1999. The role of fire in European Mediterranean ecosystems. In: Chuvieco, E. (Ed.), Remote Sensing of Large Wildfires. Springer-Verlag, Berlin, pp. 3−16.

Raison, R.J., O'Connell, A.M., Khanna, P.K., Keith, H., 1993a. Effects of repeated fires on nitrogen and phosphorus budget and cycling processes in forest ecosystems. In: Trabaud, L., Prodon, R. (Eds.), Fire in Mediterranean Ecosystems. Report no. 5 Ecosystem Research Report Series. CEE, Brussels, Belgium, pp. 347−363,.

Raison, R.J., Jacobsen, K.L., Connell, M.J., Khanna, P.K., Keith, H., Smith, S.J., Piotrowski, P., 1993b. Nutrient cycling and tree nutrition. In: Collaborative Research in Regrowth Forest in East Gippsland between CSIRO and the Victorian Department of Conservation and Natural Resources. Second Progress Report. CSIRO Division of Forestry, Canberra, Australia, pp. 8−89.

Raison, R.J., Khanna, P.K., Jacobsen, K.L.S., Romanya, J., Serrasolses, I., 2009. Effects of fire on forest nutrient cycles. In: Cerdá, A., Robichaud, P.R. (Eds.), Fire Effects on Soils and Restoration Strategies. Science Publishers, Enfield, NH, USA, pp. 225−256.

Raison, R.J., Khanna, P.K., Woods, P.V., 1985a. Mechanisms of element transfer to the atmosphere during vegetation fires. Can. J. Forest Res. 15, 132−140.

Raison, R.J., Khanna, P.K., Woods, P.V., 1985b. Transfer of elements to the atmosphere during low intensity prescribed fires in three Australian subalpine eucalypt forests. Can. J. Forest Res. 15, 657−664.

Randerson, J.T., Liu, H., Flanner, M.G., Chambers, S.D., Jin, Y., Hess, P.G., Pfister, G., Mack, M.C., Treseder, K.K., Welp, L.R., Chapin, F.S., Harden, J.W., Goulden, M.L., Lyons, E., Neff, J.C., Schuur, E.A., Zender, C.S., 2006. Science 314, 1130−1132.

Rein, G., Cleaver, N., Ashton, C., Pironi, P., Torero, J.L., 2008. The severity of smouldering peat fires and damage to the forest soil. Catena 74, 304−309.

Reyes, O., Casal, M., 2008. Regeneration models and plant regenerative types related to the intensity of fire in Atlantic shrubland and woodland species. J. Veg. Sci. 19, 575−583.

Robichaud, P.R., 2009. Post-fire stabilization and rehabilitation. In: Cerdá, A., Robichaud, P.R. (Eds.), Fire Effects on Soils and Restoration Strategies. Science Publishers, Enfield, NH, USA, pp. 299−320.

Rojo, L., Bautista, S., Orr, B.J., Vallejo, V.R., Cortina, J., Derak, M., 2012. Prevention and restoration actions to combat desertification. An integrated assessment: the PRACTICE project. Sécheresse 23, 219−226.

Ryan, K.C., Opperman, T.S., 2013. LANDFIRE − a national vegetation/fuels data base for use in fuels treatment, restoration, and suppression planning. Forest Ecol. Manage. 294, 208−216.

Schoennagel, T., Smithwick, E.A.H., Turner, M.G., 2008. Landscape heterogeneity following large fires: insights from Yellowstone National Park, USA. Int. J. Wildland Fire 17, 742−753.

Schröter, D., Cramer, W., Leemans, R., Prentice, I.C., Araújo, M.B., Arnell, N.W., Bondeau, A., Bugmann, H., Carter, T.R., Gracia, C.A., de la Vega-Leinert, A.C., Erhard, M., Ewert, F., Glendining, M., House, J.I., Kankaanpää, S., Klein, R.J.T., Lavorel, S., Lindner, M., Metzger, M.J., Meyer, J., Mitchell, T.D., Reginster, I., Rounsevell, M., Sabaté, S., Sitch, S., Smith, B., Smith, J., Smith, P., Sykes, M.T., Thonicke, K., Thuiller, W., Tuck, G., Zaehle, S.,

Zierl, B., 2005. Ecosystem service Supply and vulnerability to global change in Europe. Science 310, 1333–1337.

Scott, D.F., Curran, M.P., 2009. Soil erosion after forest fire. In: Cerdá, A., Robichaud, P.R. (Eds.), Fire Effects on Soils and Restoration Strategies. Science Publishers, Enfield, NH, USA, pp. 177–196.

Serrasolsas, I., Khanna, P.K., 1995. Changes in heated and autoclaved forest soils of S.E. Australia. II. Phosphorus and phosphatase activity. Biogeochemistry 29, 25–41.

Serrasolses, I., Vallejo, V.R., 1999. Soil fertility after fire and clear-cutting. In: Rodà, F., Retana, J., Gracia, C.A., Bellot, J. (Eds.), Ecology of Mediterranean Evergreen Oak Forests. Springer-Verlag, Berlin, pp. 315–328.

Shakesby, R.A., 2011. Post-wildfire soil erosion in the Mediterranean: review and future research directions. Earth Sci. Rev. 105, 71–100.

Shlisky, A., Waugh, J., Gonzales, P., Gonzalez, M., Manta, M., Santoso, H., Alvarado, E., Ainuddin Nuruddin, A., Rodriguez-Trejo, D.A., Swaty, R., Schmidt, D., Kaufmann, M., Myers, R., Alencar, A., Kearns, F., Johnson, D., Smith, J., Zollner, D., Fulks, W., 2007. Fire, Ecosystems and People: Threats and Strategies for Global Biodiversity Conservation. GFI Technical Report 2007-2. The Nature Conservancy, Arlington, VA, USA.

Specht, R.L., Moll, E.J., 1983. Mediterranean-type heathlands and sclerophyllous shrublands of the world: an overview. In: Kruger, F.J., Mitchell, D.J., Jarvis, J.V.M. (Eds.), Mediterranean-type Ecosystems: The Role of Nutrients. Springer-Verlag, Berlin, pp. 41–65.

Stephens, S.L., Millar, C.I., Collins, B.M., 2010. Operational approaches to managing forests of the future in Mediterranean regions within a context of changing climates. Environ. Res. Lett. 5. http://dx.doi.org/10.1088/1748-0326/5/2/024003.

Trabaud, L., 1990. Fire resistance of *Quercus coccifera* L. garrigue. In: Goldamer, J.G., Jenkins, M.G. (Eds.), Fire in Ecosystems Dynamics. Mediterranean and Northern Perspectives. SPB Academic Publishing bv, The Hague, The Netherlands, pp. 21–32.

Turner, M.G., Hargrove, W.W., Gardner, R.H., Romme, W.H., 1994. Effects of fire on landscape heterogeneity in Yellowstone National Park, Wyoming. J. Veg. Sci. 5, 731–742.

Valdecantos, A., Cortina, J., Vallejo, V.R., 2011. Differential field response of two Mediterranean tree species to inputs of sewage sludge at the seedling stage. Ecol. Eng. 37, 1350–1359.

Vallejo, V.R., Alloza, J.A., 1998. The restoration of burned lands: the case of eastern Spain. In: Moreno, J.M. (Ed.), Large Forest Fires. Backhuys Publ., Lieden, pp. 91–108.

Vallejo, V.R., Alloza, J.A., 2012. Post-fire management in the Mediterranean Basin. Israel J. Ecol. Evol. 58, 251–264.

Vallejo, V.R., Arianoutsou, M., Moreira, F., 2012a. Fire ecology and post-fire restoration approaches in Southern European forest types. In: Moreira, F., Arianoutsou, M., Corona, P., De Las Heras, J. (Eds.), Post-fire Management and Restoration of Southern European Forests. Springer, Dordrecht, The Netherlands, pp. 93–119.

Vallejo, V.R., Bautista, S., Cortina, J., 1999. Restoration for soil protection after disturbances. In: Trabaud, L. (Ed.), Life and Environment in the Mediterranean. WIT Press, Southampton, UK, pp. 301–343.

Vallejo, V.R., Serrasolses, I., Alloza, J.A., Baeza, J., Bladé, C., Chirino, E., Duguy, B., Fuentes, D., Pausas, J.G., Valdecantos, A., Vilagrosa, A., 2009. Long-term restoration strategies and techniques. In: Cerdá, A., Robichaud, P.R. (Eds.), Fire Effects on Soils and Restoration Strategies. Science Publishers, Enfield, NH, USA, pp. 373–398.

Vallejo, V.R., Smanis, A., Chirino, E., Fuentes, D., Valdecantos, A., Vilagrosa, A., 2012b. Perspectives in dryland restoration: approaches for climate change adaptation. New Forests 43, 561–579.

Walker, J., Raison, R.J., Khanna, P.K., 1986. Fire in Australian soils. In: Russell, J.S., Isbell, J.S. (Eds.), Australian Soils: The Human Impact. Univ. Queensland Press, Queensland, Australia, pp. 185–216.

Williams, J., Albright, D., Hoffmann, A.A., Eritsov, A., Moore, P.F., Morais, J.C.M., Leonard, M., San Miguel-Ayanz, J., Xanthopoulos, G., van Lierop, P., 2011. Findings and implications from a coarse-scale global assessment of recent selected mega-fires. In: International Wildland Fire Conference (5th, 9–13 May 2011, Sun City, South Africa). FAO.

Williams, B.J., Song, B., Williams, T.M., 2013. Visualizing mega-fires in the past: a case study of the 1894 Hinckley fire, east-central Minnesota, USA. Forest Ecol. Manage. 294, 107–119.

Chapter 13

Ensuring That We Can See the Wood and the Trees: Growing the Capacity for Ecological wildfire Risk Management

Douglas Paton
School of Medicine (Psychology), University of Tasmania, Launceston, Tasmania, Australia

Petra T. Buergelt
Charles Darwin University, School of Psychological & Clinical Sciences, Darwin, Australia, University of Western Australia, Centre for Social Impact and Oceans Institute, University of Western Australia, Australia & Joint Centre for Disaster Research, Massey University, Mt Cook, Wellington, New Zealand

Michael Flannigan
Faculty of Forestry, University of Toronto, Toronto, ON, Canada

ABSTRACT

This chapter discusses the need to view wildfire risk management as an activity that is embedded in the relationships that exist between people and forest environments. It outlines the issues that need to be accommodated to develop holistic approaches to wildfire risk management that can facilitate more harmonious ways of social coexistence with forest environments. This chapter discusses the causes of wildfire, how causes will change over time, the consequences that need to be predicted and managed, and the need to develop risk management strategies that accommodate the perspectives of all stakeholders in a context in which social (e.g., social development in wildland–urban interface areas) and environmental (e.g., climate change) contributions are creating a progressively more complex risk scape. How this can be accomplished is discussed in the context of the lessons that can be learned from integrating interdisciplinary perspectives on the causes, consequences, and systematic management of wildfires.

13.1 INTRODUCTION

Forests provide vital ecosystem services (e.g., clean air and water, biodiversity, carbon storage, nontimer forest products, and recreation) that are integral to culture, furnish people and communities with important economic resources, sustain well-being (Clayton and Opotow, 2003), and provide lifestyle amenities

Wildfire Hazards, Risks, and Disasters. http://dx.doi.org/10.1016/B978-0-12-410434-1.00013-0

that are enjoyed by many. Over time, specific ecosystems have symbiotically coevolved in interaction with wildfires. Wildfires play an integral part in shaping and maintaining the ecology of forests, creating wildfire-adapted and -dependent ecosystems, and are an essential tool for landscape management (see Chapters 2, 3, 7, and 9). However, because forests can periodically turn hazardous, wildfire is a natural and inevitable part of forest landscapes around the world. The high degree of interdependence that exists between people and forests means that forest and fire can have both beneficial and detrimental consequences for people and communities (see Chapters 5, 6, 8, and 9).

This book addressed those circumstances in which forests become part of a societal hazardscape. This occurs when social (e.g., people's decisions about where to live and develop, societal approaches to mitigation) and ecological (e.g., vegetation type and climate) characteristics interact in ways that threaten or result in losses to what people value. If people and societies wish to continue to enjoy the social, cultural, and economic amenities afforded by forests and forest landscapes, they need to develop ways of coexisting harmoniously with them.

The fact that harmonious coexistence is possible is evident from the contributions to this text that attest to the long history of doing so in Aboriginal societies in which forest and fire were integral to community lifecycles. Aboriginal people in America, Canada, Europe, Australia, and India have been intentionally, actively, and intuitively using and managing fire dating back as far as 0.6 million years before present. The discussion in Chapters 2, 3, 5–7 paints a picture of how traditional and indigenous knowledge and skills were build up and transmitted over thousands of years in ways that enabled our ancestors to live in harmony with fire. These contributors also illustrate how the disregard of thousands of years of firewise forest and fire management led to the loss of pertinent community knowledge, skill, and practices and ultimately to the contemporary crisis in wildfire management. This loss reflects historical trends in, for example, land use and lifestyle choices being made progressively from anthropocentric, rather than from social–ecological, perspectives.

The past 200 years have witnessed the balance between people's perception of and the use of fire as a tool and resource versus fire as a hazard shifting progressively toward the hazard end of the continuum. Factors such as the colonial importation of inappropriate land use and fire management practices related to the invasion of forest to create permanent settlements or agricultural land progressively led to fire being seen as a hazard that has to be managed (see Chapters 2, 3, 5–7, and 9). Rather than perceiving fire as a creative force and respecting and appreciating fire as aboriginal people had done, colonizers perceived fire as a destructive force that had to be controlled. These factors have resulted in populations losing much of the fire management knowledge and skill acquired over centuries of understanding the benefits of coexisting with forests and replacing them with reactive management practices such as fire suppression. The practice of suppressing wildfires to protect the resources people and societies created has, ironically, resulted in the dominant aspects of wildfire management often creating environments that increased the frequency and intensity of wildfires. Thus, a significant challenge for wildfire risk management is to regain mastery in using and managing fire and developing more sustainable, harmonious ways of coexisting with essential forest habitats in ways that accommodate both beneficial and hazardous circumstances.

The contents of this volume illustrate the growing awareness that the challenges wildfire risk management pose are increasingly "wicked problems"—problems that are complex and characterized by a wealth of diverse but interconnected relationships and interdependencies (see Chapters 2, 5, and 7). That is, ecological and environmental (e.g., vegetation, topographical, and meteorological phenomena), psychological, personal, historical, natural/physical, religious/spiritual, social/cultural, technological, economic, legal, and political dimensions that interact across different scales (e.g., household, neighborhood, community, shire/municipality, state, federal, global) interact to create both the context in which wildfire risk develops and in which it needs to be managed. This means that the effectiveness of wildfire risk management becomes a function of how well expertise drawn from these diverse areas is integrated into a comprehensive wildfire risk management strategy. However, the complexity of facilitating this integration is complicated by the fact that much of what we know has emerged from diverse disciplinary "trees" of knowledge.

What is needed is a way of ensuring that the extensive intellectual investment in understanding individual trees of knowledge does not obscure our appreciation of the fact that the trees are all in the same risk management wood. The comprehensive understanding of wildfire risk and its effective management will only arise from understanding how the trees form part of the wood and that all stakeholders are making complementary contributions to sustainable "woodland" management.

How we ensure that we can see the wood and the trees starts with volumes such as this one. This volume presents not only the diverse issues that need to be understood but it also, hopefully, lays a foundation for the crossfertilization of knowledge and sows the seeds of interdisciplinary collaboration to build an ecology of wildfire risk and its management. To capture all these elements, and the interactions among and between them, a comprehensive ecological perspective is required to ensure that the benefits of the knowledge discussed can be fully realized (e.g., Buergelt and Paton, in press). Applying the ecological perspective to wildfires and their management will enable the identification of the most effective and sustainable risk management levers, and contribute to creating innovative solutions that enable us to effectively manage wildfires and to balance environmental, social, economic concerns and living with wildfires in a sustainable and dynamic equilibrium. To do so, it is necessary to understand the causes of wildfire, how causes will change over time, the consequences that need to be predicted and managed, and how interdisciplinary perspectives on these can be pressed into service to managing wildfire risk. Discussion commences with the causes and how they are likely to change.

13.2 CAUSES AND CONSEQUENCES

13.2.1 Causes

Wildfire is a common occurrence globally, with recent estimates of annual area burned ranging from 273 to 567 Mha, with an average of 383 Mha (Schultz et al., 2008). Fire activity over a region is strongly influenced by four factors: fuels, climate—weather, ignition agents, and people (Flannigan et al., 2005). Fuel amount, type, continuity, structure, and moisture content are critical elements for fire occurrence and spread. Fuel continuity is required for fires to spread; some suggest that at least 30% of the landscape needs to have fuel for a fire to spread. This is important in many drier parts of the world where precipitation is required preceding the fire season for the growth of vegetation and the subsequent generation of surface fuels available to carry fire on the landscape (Swetnam and Betancourt, 1998; Meyn et al., 2007).

Fuel structure can also be important in fire dynamics, for example, under-story trees and shrubs in a forest can act as ladder fuels that help a surface fire to reach the tree crowns and thereby generate a faster moving and much more intense fire. Although the amount of fuel, or fuel load, and fuel distribution (vertical and horizontal) affect fire activity, fuel moisture largely determines fire behavior, and has been found to be an important factor in the amount of area burned. There are two common mechanisms for wildfire ignition: people and lightning. Estimates are that 90% of wildfires are started by people.

Weather and climate—including temperature, precipitation, wind, and atmospheric moisture—are critical aspects of fire activity. Weather is a key factor in its own right, but it also influences fuel and ignitions. Fuel moisture, which may be the most important aspect of fuel is a function of the weather, and weather and climate also in part determine the type and amount of vegetation (fuel) at any given location. Additionally, lightning as one of the two main causes for wildfire is largely determined by the meteorological conditions. Weather arguably is the best predictor of regional fire activity for time periods of a month or longer. For example, Cary et al. (2006) found that weather and climate best explained modeled area burned estimated from landscape fire models compared with variation in terrain and fuel pattern. Although wind speed may be the primary meteorological factor affecting fire growth of an individual fire, numerous studies suggest that temperature is the most important variable affecting overall annual wildfire activity, with warmer temperatures leading to increased fire activity (Gillett et al., 2004; Flannigan et al., 2005; Parisien et al., 2011).

The reason for the positive relationship between temperature and regional wildfire is threefold. First, warmer temperatures will increase evapotranspiration, as the ability for the atmosphere to hold moisture increases rapidly with higher temperatures, thereby lowering water table position and decreasing forest floor and dead fuel moisture content unless there are significant increases in precipitation. Second, warmer temperatures translate into more lightning activity that generally leads to increased ignitions (Price and Rind, 1994). Lastly, warmer temperatures may lead to a lengthening of the fire season (Wotton and Flannigan, 1993; Westerling et al., 2006). While testing the sensitivity of landscape fire models to climate change and other factors, Cary et al. (2006) found that area burned increased with higher temperatures. This increase was present even when precipitation was increased, although the increase in area burned was the greatest for the warmer and drier scenario. The bottom line is that we expect more fire in a warmer world.

The global climate is warming, and this may have a profound and immediate impact on wildfire activity. If fire weather and climate are critical aspects of fire activity, then it appears that the future will be more conducive to fire overall (Flannigan et al., 2009). Some suggest that wildfire activity has already increased due to climate change. For example, Gillett et al. (2004) suggest that the increase in area burned in Canada during 1974–2014 is due to human-caused increases in temperatures. These increases were occurring despite increasing fire suppression effectiveness and increased coverage by fire suppression resources. Wildfires and climate change are amplifying each other. The vast areas that burn worldwide annually are creating products that serve to amplify climate change. Climate change, in turn, increases the frequency and severity of wildfires (see Chapters 2, 3, 5, and 7) and thus the consequences they will create for those at risk.

13.2.2 Consequences

While wildfires constitute a natural process, the growth in the risk fire poses arises from ecological changes created by decades of fire suppression, changes in land-use patterns, increasing social complexity created by the growing number and diversity of organizations and people living in areas susceptible to wildfires, and climate change. With the exception of Antarctica, these processes operate worldwide. No continent is immune to wildfire risk. Further, the reach of this hazard is extending. Chapter 5 discussed how fire risk is spreading northward throughout Europe. Chapter 8 discussed how fire, as a hazard, has emerged as a growing hazard in Taiwan. Elsewhere, the specter of more frequent, more intense, and longer duration fires is a source of growing concern. Furthermore, the consequences they create can extend beyond the immediate impact of the fires.

For example, large wildfires produce impacts globally through atmospheric emissions of greenhouse gases and particles, and locally through impacts on air quality, ecosystems, landscapes, and human structures (see Chapters 6, 10, and 12). The air pollution caused by biomass burning in rural areas in India, for instance, is causing bronchitis and asthma not only in people who live in close proximity to the source of fire but also for those living much further afield (see Chapter 9). This air pollution might be the cause of hundreds of thousands of deaths each year mainly from lung and heart diseases. The fact that the associated risk need not be solely a regional or national issue was further highlighted in Chapters 6 and 10 when they, respectively, discussed how smoke hazards from Indonesia transcend national borders to create regional impacts and how large, difficult to manage Russian fires make substantial contributions to global emissions.

At local and regional levels, wildfires have an impact on all dimensions of the ecological framework: individual, natural/physical, religious/spiritual, historical/cultural, social, technological, economic, and political dimensions. Potential losses can impact several domains of social and community life. In addition to their potential to create a loss of life, wildfires can result in the loss of historically significant heritage sites and artifacts in ways that can affect people's place attachment and sense of belonging for community members (see Chapter 9). The destruction of forests, farms, and livestock can destroy the livelihoods of people who depend on forests and live off the land. Wildfire experience also triggers disruption of family routines, loss of jobs, and volunteer work, and leads to higher rates of crime and violence (see Chapters 3 and 6).

Wildfires hazards also degrade air, water, and soil quality and quantity— the very resources our lives depend on (see Chapter 9). Wildfires have an impact on soils and water quality (see Chapter 12), contribute to soil erosion and top soil losses and can reduce soil fertility and nutrient content. Changes to soil and water quality can also contribute to food and water insecurities. In spite of net nutrient ecosystem losses, burned ecosystems are often rich in available nutrients right after the fire in the short term. However, soil microbial biomass and diversity may take long time to recover, and soil formation is an extremely slow natural process. Hazardous fires can be the indirect cause of landslides, mudslides, erosion, increased water run-off, flash floods, and soil depletion.

In the built environment, wildfires destroy or damage thousands of homes, businesses, public buildings (e.g., churches, stores, and libraries), and roads. Infrastructure loss impacts the economy. Economically, wildfires can result in a wide variety of substantial direct and indirect short- and long-term costs to in-dividuals (e.g., loss of health/well-being that require treatment or hinder working, houses, possessions, farms, businesses, and jobs); communities (e.g., loss of residents, loss of jobs and businesses, loss of revenue, loss of community assets and recovery, and reconstruction issues); local, state, and federal governments (e.g., preparedness, response, and assistance for individuals and communities); insurance (e.g., insurance assessment and payouts), and businesses (e.g., loss of business buildings and material, loss of jobs, and loss of business information). In rural communities, the loss can be protracted if businesses remain closed for prolonged periods of time (e.g., from lack of business continuity planning, loss of customer or client base if people have to move out of an area (Paton and McClure, 2013)). Rural fire-affected areas are also sensitive to losses from postfire declines in tourism.

Socially, the destruction that fire wreaks on the natural and built environment can result in thousands of people having to evacuate (see Chapters 2, 3, and 7). Most evacuations are short term, but some create a need for long-term resettlement, including when people migrate permanently due to fear of the recurrence of wildfires. In poorer countries that cannot afford providing evacuation centers or emergency housing, wildfires result in a large number of people being homeless.

On the positive side, the sense of shared fate created by wildfires can unite residents and act as a catalyst for their experiencing community spirit and working together as communities (McAllan et al., 2011; see Chapter 3). In general, more research is needed into positive/salutogenic aspects of wildfires in all dimensions (e.g., creating a sense of community after fires). A knowledge of how such outcomes can arise phoenix-like from the ashes of wildfire events would give us valuable clues as to how to create and facilitate their emergence and optimize recovery and growth opportunities in a postfire environment and complement efforts on ecological restoration (see Chapter 12). The finding that the causes of many, if not most, wildfires can be attributed to social and personal decisions and actions is both a cause for concern, particularly with regard to arson, and a context in the social, psychological, and cultural sources of social wildfire risk management can be sourced (see Chapters 2 and 5). The empirical bases for such intervention are known (e.g., Frandsen et al., 2012; McFarlane et al., 2011; Paton et al., 2008). The next step is to develop them within appropriate social and cultural contexts. The starting point for such an endeavor is learning the lessons from research and experience and embedding them in future ecological wildfire risk management strategies.

13.3 LESSONS

13.3.1 Historical, Spiritual, Religious, Social, and Cultural Lessons

Historical lessons that can contribute to the development of sustainable risk management practices can be sourced from aboriginal ecological knowledge and practices and from recovering traditional ecological knowledge and practices (see Chapters 2, 3, 7, and 9). Some countries reestablished the traditional use of fire to manage the land and to reduce fuel hazard (see Chapter 5). Yet, the more extensive and widespread losses that the specter of climate change (e.g., the growing risk of megafires) introduces into this context is altering local conditions, making it difficult to restore and apply traditional fire regimes.

The contribution of spiritual and religious characteristics to wildfire risk and how it is managed is a little-studied area. However, its appearance on future research agendas is justified by the fact that several contributors (e.g., Chapters 2, 6, and 9) identified these as important factors. The work discussed in these chapters illustrates some of the benefits that can accrue from increasing our understanding of the spiritual and religious dimensions of risk and for how social risk management needs to be conceptualized to accommodate these dimensions in risk management.

In the social domain, research on prefire social dynamics provides the foundation for facilitating public support for mitigation (e.g., prescribed burning), encouraging sustained household and community readiness, developing evacuation plans, facilitating understanding and adoption of alternative social response strategies, such as the "Shelter in Place" and the "Stay and Defend or Leave Early" elements of a social risk management strategy. This work also underpins the extension of these efforts into exploring how social dynamics influence postfire response and recovery (see Chapter 2).

The latter work illustrates one link between the social and technological aspects of risk management. In particular, it illustrates how the return on investment in warning information and technologies (see Chapter 11) can only be realized if social risk communication and community engagement strategies facilitate the capacity of people to attend to warnings and act on them in timely and appropriate ways. This cannot be assumed (e.g., Prior and Paton, 2008). This provides a good example of what is meant by a social—ecological relationship; the warnings processes that provide one link between people and environment need to be developed simultaneously with the social capability to act on them. Thus, social intervention may need to be implemented alongside the development of technological aids to wildfire risk management.

13.3.2 Technical Lessons

Technological advances provide new and improved tools for recording, processing, and integrating increasingly more data and data from many different areas more accurately (see Chapter 4). This increases the quality and speed in which we can forecast, detect, and evaluate wildfire risk; warn citizens; determine the geometric shape of fire expansion; calculate the economic costs of actions; allocate resources and fire fighting methods; calculate fire protection priorities; and conduct computer simulations of wildfires. To apply the numerous technical options available to the wildfire management area, new research projects need to be conducted (see Chapter 4).

Effective fire danger rating and early warning systems are crucial for a broad spectrum of wildfire management decision-making wildfire management (see Chapter 11). They can operate at global to local levels. Fire danger rating has become the cornerstone of national fire management programs. Worldwide, there are many different national fire danger rating systems in use, and some systems serve several countries (see Chapter 11).

In Chapter 11, de Groot et al. predict that technological enhancements in fire danger rating and early warning information can be expected. For example, spatial mapping of precipitation (using radar and satellite data) would significantly improve spatial accuracy of fire danger maps. Remote sensing would enable frequent updates on fuel structure and load information, which would enhance the quality and speed of weather-driven fire behavior models. Remote monitoring of live fuel moisture will assist in determining landscape-level fuel flammability and seasonal criteria that are important to predicting fire behavior. A related challenge stems from the need to develop more future-oriented warnings to trigger actions to confront the climate change implications for societal risk. This challenge encompasses issues ranging from environmental prediction to reconciling scientific views (e.g., scientific disagreement is interpreted by citizens as a reason for inaction (e.g., Paton and McClure, 2013)) to developing an understanding on how to mobilize community dynamics to facilitate large-scale, sustained community change.

Social intervention is required to fully realize the benefits of technology derived from strategies that include interactive communication and engagement/outreach between agencies and citizens and that progressively develop community capability (see Chapter 2). Recognizing linkages among and between the various dimensions can help players involved to more effectively cocreate effective bushfire management policies and practices. Facing up to the social risk management challenges posed by climate change means that additional factors will have to be accommodated in response to changes on the risk scape that are gradual (e.g., small incremental increases in temperature) and less readily discerned from personal and social perspectives. This is particularly important as individuals readily adapt and create a "new normal" when experiencing change in small increments, and this acts to constrain their capacity to appreciate significant qualitative changes in crucial circumstances, and lessen any perceived urgency about taking action to manage risk (e.g., Tickell, 1990). In contrast, if people were dropped suddenly into an environment characterized by megafires, they would recognize a significant change, and one demanding immediate action (this is colloquially referred to as the "boiling frog syndrome"). This element must be included in risk communication and outreach. However, as Buergelt and Smith (see Chapter 7) point out, mounting research shows that education has not made the desired difference. This reflects the failure to accommodate empirical research into the development of risk communication guidelines.

13.3.3 Postfire Restoration Lessons

More attention needs to be directed to postforest-fire restoration work (see Chapters 8 and 12). Carefully prioritized, and scientifically and technically justified, postfire restoration of burned landscapes reduces damages and suppression costs in the long term (see Chapter 12). Postfire restoration needs to be in alignment with the specific local ecological and social conditions and risks, with fire impacts on ecosystem fire resilience, the long-term management objectives. Chapter 2 introduced the need for more attention to be directed to postfire social dynamics and their role in facilitating social recovery. Efforts in this direction can build on work on investigating community experiences of recovery from large-scale wildfires (e.g., McAllan et al., 2011; Paton et al., 2014).

13.3.4 Wildfire Management

Several lessons for wildfire risk management can be discerned in the text, and particularly for countries transitioning from old to new fire management practices (i.e., total suppression and prevention versus preparing and community engagement) and developing risk management strategies that balance these different goals to facilitate comprehensive risk management. The importance of effectively managing this shift can be traced to the growing complexity arising from, for example, interaction between social migration into wildland—urban interface areas and growing environmental sources of risk from climate change (e.g., longer fire seasons, more intense fires, and the threat from megafires). Further complexity is introduced into this process as a result of the fact that comprehensive risk management calls for a diverse range of stakeholders to be included in planning and implementation.

The collective knowledge, skill, and commitment required to understand and manage wildfire risk exists within several disciplines, professions, and agencies, making identifying relevant stakeholders and facilitating collaboration between these key players an important task. Stakeholders include agencies with fire management responsibility, forest owners (e.g., federal, state, municipalities, and monasteries, private), and community members and institutions. Future research is needed to investigate the practical, social, legal, economic, and political implications of reconciling the needs and goals of such a diverse group into coherent management plans and activities. It will also be pertinent to devote research resources to better understand the community role in this process and to accommodate the social and cultural diversity they introduce into the risk management mix. Some ideas about how this can be done were discussed in Chapters 2, 3, and 6 (see also Frandsen et al., 2012).

A significant challenge to risk management is thus how to develop a process that recognizes diverse inputs and outputs and that ensure that all stakeholders play complementary roles in the process. While fire risk represents a shared context in which these stakeholders ostensibly come together, recognition of this risk is not enough to ensure collaborative action. Developing complementary collaboration first requires identifying how to bring diverse stakeholder groups together to form a superordinate group (built around shared risk and the need for collective input and action to manage that risk) whose respective contributions complement one another under a comprehensive risk management umbrella (Paton and McClure, 2013). Achieving this outcome may require drawing on business and organizational psychology expertise in group and team development. For example, facilitating the development of effective and sustained superordinate risk management relationships could be pursued by using scenario planning, a technique designed to channel the input from diverse groups into ways of managing future best- and worst-case scenarios in risk management contexts (e.g., Moore et al., 2013). Techniques such as these can also inform the development of sustainable wildfire mitigation practices (e.g., prescribed burning) that rely on concerted action from diverse stakeholders for their effective implementation.

13.3.5 Prescribed Burning and Fire-Use Practices

The historical debate regarding the management of vegetation to reduce fire hazards and the management of the vegetation for conservation is continuing. Using fire appropriately is one of the most complex and divisive issues within wildfire management (see Chapter 7), particularly when diverse public perceptions of and/or support for prescribed burning can lead to significant community conflicts (see Chapters 3 and 7). Such conflicts are being fuelled by the growing number of people migrating into wildland–urban interface areas (see Chapter 7 and Radeloff et al., 2005). Negative public perceptions, restrictive legal frameworks, complex territorial structures, and lack of experience among professionals restrict the use of prescribed burning (see Chapter 5). Chapter 3 discusses how Canada is addressing such issues by integrating an ecocultural focus with mechanical thinning. Encouraging professionals and local community collaboration in the planning and conducting prescribed burning increases the number of people skilled in using fire in sustainable ways and diffuses conflicts (see Chapter 5) and increases the protection offered to communities and natural environments alike (see Chapter 7). It is becoming increasingly apparent that it is crucial to find the right balance: too much fire can alter the composition and structure of the natural ecosystem; too little can predispose the natural ecosystem to larger and more severe wildfires. Climate change pressure will amplify these problems (e.g., rainfall decline resulting in less spring weather suitable for prescribed burning), increasing the urgency inherent in the pursuit of sustained and balanced mitigation practices. Irrespective of the element of the risk management strategy being considered, it is important to appreciate that it will at some stage rely on government support.

13.3.6 Government and Governance

The causes of fire, and how the various components of risk are managed, are shaped by government policies and how political and economic imperatives can often override other considerations (see Chapters 5 and 6). In all countries included in this book, wildfire management is fundamentally a federal responsibility: government provides leadership; manages, facilitates, and co-ordinates actions; builds capabilities; and designs and reinforces policies and laws (see Chapters 2, 5−7). While current wildfire policies still favor suppression over prevention, successful changes have been implemented by some European countries; the United States, Canada, and Australia (see Chapters 2, 3, 5, and 7). These countries are aiming to create holistic, cohesive, and comprehensive federal wildfire management policies and strategies that cover both prevention and suppression through central planning and decentralized operations (at local-, state-, and federal-level solutions using collaborative process that bring together the diverse players from all levels—see Chapter 2).

If it is to be implemented effectively, this collaborative process requires greater intergovernmental coordination in prefire preparations during-fire management and postfire recovery planning and operations (see Chapter 2). It also requires creating a shared vision for comprehensive wildfire risk management by all stakeholders (see Chapter 3), and one that encompasses, for example, reducing fuel, empowering communities, developing their ca-pabilities, and facilitating sustained collaboration among key stakeholders (see Chapters 2 and 7). The policy to practice transition inherent in this approach is another area that could benefit from professionally facilitated techniques such as scenario planning to manage conflict associated with the structural, prac-tical, and social−psychological differences mobilized in multijurisdictional contexts to create sustained collaboration, ecosystem restoration, and response and recovery policies and practices (see Chapter 7).

However, in general, current policies and practices are not only progres-sively struggling and proving to be inadequate for meeting the challenges ahead but they often also increase risk and become unsustainable (see Chapter 7). Chapter 7 discusses how considerable resources have been invested in assessing what did not work in previous wildfires, but this has not been matched by resourcing learning from and implementing lessons learned from previous wildfire. Buergelt and Smith also discuss how the effectiveness of education and training of bushfire risk management is compromised by the diverse players playing largely in isolation rather than collaborating and coordinating to leverage their impact. This suggests that exploring how regu-latory and legislative intervention could make further contributions to pro-moting sustained interagency learning and training.

13.3.7 Businesses as Stakeholders

The consequences wildfire can have for business and economic activity identify the need for business and commercial companies to be stakeholders in wildfire risk management. Businesses and communities play complementary roles in risk management. The former provide employment. The latter are employees and customers. Unless both business and communities prepare, the important complementary roles they play in community vitality will be lost, risk will be amplified, and recovery slowed. Business readiness can sustain their bottom lines during periods of crisis and, by acting with a sense of social responsibility, actively contribute to the communities they are located in (Paton and McClure, 2013). However, the development of this complementary capability needs to be planned for in both business continuity planning and in community planning.

13.4 PATHWAYS FORWARD

Wicked problems can only be solved, and people can only cope with rapid change if the majority of people in organizations are learning fast, understanding complexity, synthesizing seemingly separate pockets of knowledge, and are able to apply this new knowledge to create innovative solutions and products/services (see Chapter 7). This requires transforming people's mindset using cutting edge learning and teaching approaches (e.g., transformative education, accelerated and transformative learning) to equip stakeholders with physical, mental, emotional, and spiritual capabilities that have them accomplish quantum leaps in performance and working in teams (see Chapter 7).

It is crucial to conceptualize wildfire risk holistically. What this might look like is summarized in Figure 13.1. Central to this conceptualization is appreciating that societal- and community-level resources play complementary roles in this process (Figure 13.1). Just as societal-level input must be conceived as comprising interdependent contributions from different agencies, so too must the community contribution be defined in a way that acknowledges the diversity introduced by the existence of several (relational) communities (depicted by the separate circles in the community component of Figure 13.1). Societal and community levels of analysis interact directly and indirectly with forest environments (e.g., via the action of, e.g., land management agencies for the former, and lifestyle and employment interests for the latter). The relationship of both with forest environments is also mediated by risk management and warning processes. Figure 13.1 also illustrates the enveloping role that climate change processes will play in how social and technical processes develop and function over time.

To create sound, effective fire management, it is vital to increase collaboration among and between key players, to design appropriate institutional structures, perceive communities as valuable allies, to integrate traditional or local knowledge, and to ensure that land use is based on scientific data (see Chapter 5). Creating fire-adapted communities (see Chapter 2) capable of living in harmony with wildfires requires the proactive participation, collaboration, and cooperation across stakeholders, dimensions, geography, and scales of analysis (e.g., regional, global) (see Chapters 3 and 5) to optimize the use of knowledge and resources (see Chapter 6). To pursue this objective, the expertise from disciplines such as occupational psychology, business

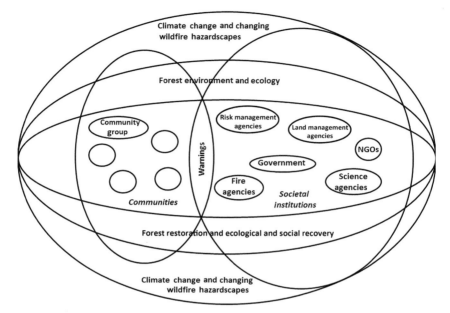

FIGURE 13.1 Ecological wildfire risk management model.

management, and sociology could be used to facilitate the organizational development and strategic staff development required to cultivate the mindset and skills necessary for interdisciplinary collaboration and to develop the interagency structures necessary for collaboration (e.g., applying sociocracy principles—see Chapter 7). The same applies to engaging communities. Community participation requires a concerted, interactive collaborative effort of communities, government and private sectors (see Chapter 6) focused on the wildfire context.

The interactions between and among diverse individual, historical, natural/physical, religious/spiritual, social, technological, economic, and political characteristics over time have not only created the current challenges to wildfire risk management but also open the door to resolving current challenges in sustainable ways. Doing so involves developing a holistic conceptualization of wildfire risk in ways in which the interdependencies between the wood and the trees are recognized and actively accommodated in wildfire risk management. Consequently, it will be important to have an accurate understanding of each of these aspects and how they interact over time, and of how the complex interactions of the dimensions are affected by, and can influence, wildfire outcomes (see Chapters 2 and 7) in ways that sustain the beneficial aspects of forest and fire and reduce their negative impacts (see Chapter 3). This involves, ideally, developing the capacity for harmonious social—ecological relationships in which stakeholders value and respect this relationship and can contribute complementary perspectives, interpretations, and actions to a wildfire risk management process that is fundamentally nested in ecological and natural environmental contexts.

Framing wildfire risk management as a social—ecological issue has typically separated the social/human contribution and the natural ecosystems contribution to wildfire risk. This has not been serving us well. It has created a disconnect between the two, perpetuated a tendency to miss salient linkages between the two (see Chapter 5), and has obscured routes to understanding their interdependence. It might serve us better if we move toward seeing people and societal systems as part of the natural ecosystem because that creates a connection with and hence a better understanding of the multifaceted, interconnected nature of wildfire risk management. Such approaches could lay the foundation for better stakeholder engagement, with each other and between them and their environment, in ways that facilitate their collaborating on mitigation, warning, readiness, recovery, and restoration efforts in relation to wildfire events whose scale, complexity, and consequences are only likely to increase in the future.

REFERENCES

Buergelt, P.T., Paton, D., 2014. An ecological risk management and capacity building model. Hum. Ecol. 42, 591—603.

Cary, G., Keane, R., Gardner, R., Lavorel, S., Flannigan, M., Davies, I., Li, C., Lenihan, J., Rupp, T., Mouillot, F., 2006. Comparison of the sensitivity of landscape-fire-succession models to variation in terrain, fuel pattern, climate and weather. Landsc. Ecol. 21, 121—137.

Clayton, S., Opotow, S., 2003. Identity and the Natural Environment: The Psychological Significance of Nature. MIT Press, Cambridge.

Flannigan, M.D., Krawchuk, M.A., de Groot, W.J., Wotton, B.M., Gowman, L.M., 2009. Implications of changing climate for global wildland fire. Int. J. Wildland Fire 18, 483—507.

Flannigan, M.D., Logan, K.A., Amiro, B.D., Skinner, W.R., Stocks, B.J., 2005. Future area burned in Canada. Clim. Change 72, 1—16.

Frandsen, M., Paton, D., Sakariassen, K., Killalea, D., 2012. Nurturing community wildfire preparedness from the ground up: evaluating a community engagement initiative. In: Paton, D., Tedim, F. (Eds.), Wildfire and Community: Facilitating Preparedness and Resilience. Charles C. Thomas, Springfield, Ill, pp. 260—280.

Gillett, N.P., Weaver, A.J., Zwiers, F.W., Flannigan, M.D., 2004. Detecting the effect of climate change on Canadian forest fires. Geophys. Res. Lett. 31, L18211. http://dx.doi.org/10.1029/2004GL020876.

McAllan, C., McAllan, V., McEntee, K., Gale, W., Taylor, D., Wood, J., et al., 2011. Lessons Learned by Community Recovery Committees of the 2009 Victorian Bushfires. Cube Management Solutions, Victoria, Australia.

McFarlane, B.L., McGee, T.K., Faulkner, H., 2011. Complexity of homeowner wildfire risk mitigation: an integration of hazard theories. Int. J. Wildland Fire. http://dx.doi.org/10.1071/WF10096.

Meyn, A., White, P.S., Buhk, C., Jentsch, A., 2007. Environmental drivers of large, infrequent wildfires: the emerging conceptual model. Phys. Geogr. 31, 287—312.

Moore, S.S., Seavy, N.E., Gerhart, M., 2013. Scenario Planning for Climate Change Adaptation: A Guide for Resource Managers. Point Blue Conservation Science and California Coastal Conservancy.

Parisien, M.-A., Parks, S.A., Krawchuk, M.A., Flannigan, M.D., Bowman, L.M., Moritz, M.A., 2011. Scale-dependent controls on the area burned in the boreal forest of Canada, 1980−2005. Ecol. Appl. 21, 789−805.

Paton, D., Buergelt, P.T., Prior, T., 2008. Living with bushfire risk: social and environmental influences on preparedness. Aust. J. Emerg. Manag. 23, 41−48.

Paton, D., Johnston, D., Mamula-Seadon, L., Kenney, C.M., 2014. Recovery and development: perspectives from New Zealand and Australia. In: Kapucu, N., Liou, K.T. (Eds.), Disaster & Development: Examining Global Issues and Cases. Springer, New York, NY.

Paton, D., McClure, J., 2013. Preparing for Disaster: Building Household and Community Capacity. Charles C. Thomas, Springfield, Ill.

Price, C., Rind, D., 1994. Possible implications of global climate change on global lightning distributions and frequencies. J. Geophys. Res. 99 (D5), 10823−10831. http://dx.doi.org/10.1029/94JD00019.

Prior, T., Paton, D., 2008. Understanding the context: the value of community engagement in bushfire risk communication and education. Observations following the East Coast Tasmania bushfires of December 2006. Australas. J. Disast. issues Stud. http://www.massey.ac.nz/~trauma/issues/2008-2/prior.htm.

Radeloff, V.C., Hammer, R.B., Stewart, S.I., Fried, J.S., Holcomb, S.S., McKeefry, J.F., 2005. The wildland-urban interface in the United States. Ecol. Appl. 15, 799−805.

Schultz, M.G., Heil, A., Hoelzemann, J.J., Spessa, A., Thonicke, K., Goldammer, J.G., Held, A.C., Pereira, J.M.C., van het Bolscher, M., 2008. Global wildland fire emissions from 1960 to 2000. Global Biogeochem. Cycles 22, GB2002. http://dx.doi.org/10.1029/2007GB003031.

Swetnam, T.W., Betancourt, J.L., 1998. Mesoscale disturbance and ecological response to decadal climatic variability in the American Southwest. J. Clim. 11, 3128−3147.

Tickell, C., 1990. Human effects of climate change. Geogr. J. 156, 325−329.

Westerling, A.L., Hidalgo, H.G., Cayan, D.R., Swetnam, T.W., 2006. Warming and earlier spring increase Western U.S. forest wildfire activity. Science 313, 940−943.

Wotton, B.M., Flannigan, M.D., 1993. Length of the fire season in a changing climate. Forest. Chron. 69, 187−192.

Index

Note: Page numbers followed by "f" and "t" indicate figures and tables respectively.